Creating Lifelong Math & Science Learners

A Creative Science and Math Activity Book for Parents

Written and Illustrated By: Mary Taylor Overton

Edited by-

Dr. Anthony Overton- Assistant Professor of Marine Estuarine Environmental Science in the Department of Biology in the Thomas Harrot College of Arts and Science, East Carolina University, Greenville, N. C.

Dr. Rhea Miles-Associate Professor in the Department of Mathematics and Science Education, East Carolina University, Greenville, N. C.

Bloomington, IN Milton Keynes, UK

authorHOUSE®

AuthorHouse™
1663 Liberty Drive,
Suite 200
Bloomington, IN 47403
www.authorhouse.com
Phone: 1-800-839-8640

AuthorHouse™ UK Ltd.
500 Avebury Boulevard
Central Milton
Keynes, MK9 2BE
www.authorhouse.co.uk
Phone: 08001974150

First published by AuthorHouse 10/29/2007

ISBN: 978-1-4208-9929-0 (sc)

Printed in the United States of America
Bloomington, Indiana

This book is printed on acid-free paper.

Dedication
Love to my children **Anthony** and **Camille**.

Thanks to my good friend **Juliet Ciardi Raines** for her encouragement.

About the Author
In 2004, Mary Taylor Overton retired from the Washington, D. C. Public Schools and now resides in Atlanta, Ga. She was an Early Childhood Education Specialist, at John Eaton Elementary School, in Washington. She is widowed and has 2 children, Anthony and Camille, both with advanced degrees in science. Mary has 4 grandchildren, Zachary, Taylor, Alex, and Justin. It was with her grandchildren in mind that **Creating Life-Long Math and Science Learners** was written.

Table of Contents

CREATING LIFE-LONG MATH AND SCIENCE LEARNERS

A Creative Science and Math Activity Book for Parents

Attention Parents of Children 2-6 Years:
This is a must use book for your child's future!

Note to Parents:

Children will learn more between the ages of birth and six years of age than at any other time in his/her life! Verbally, the child will acquire a spoken command of thousands of words, and cognitively he/she will have an understanding of hundreds more. During this time the child will have attempted and acquired many new skills, and stored a wealth of knowledge and understanding about his/her world. These are the child's developmental years when he/she is constantly searching for ways to explain and understand everyday occurrences. There is so much to see and so many mysteries to solve, such as: "What? How? Why? Where? When?"

Learning is a complex and challenging process; and it does not take place in a vacuum. Children learn when they are exposed to rich, stimulating environments, age-appropriate materials, guidance, and time. It is a parent's job to make sure that all of these components are in place for their child. Parental involvement and understanding of the learning process are the keys to any child's success.

Today's child is born into a highly competitive Math, Science, and Technology based world. Everything that the child can/cannot see, taste, feel, smell, or hear is in some way connected to Math, Science, and/or Technology. At a very early age children are exposed to this ever-changing world. Today's discoveries and knowledge are often obsolete tomorrow. It is a world where a child's success in life is directly related to the ability to understand and apply scientific, mathematic, and technical skills to his/her everyday life experiences. Understanding math and science concepts is a step-by-step intensive process that must evolve over time. The formative years are ideal for laying the foundation for strong math and science concepts as well as other related discoveries.

Creating Life-Long Math and Science Learners is an activity based book that empowers parents by providing the tools needed to give their child a competitive edge. The book provides suggestions for establishing developmental and educational math and science goals for children. The activities are aimed at helping parents to address their individual child's interests, needs, and learning style. Provided are comprehensive, engaging, challenging, and creative suggestions for using a child's natural curiosity to make math and science connections and discoveries. It has long been established that young children learn best through concrete visual and tactical interaction and experiences. **Creating Life-Long Math and Science Learners** is filled with thousands of child-centered activities that support the young child's cognitive development. The activities are age-appropriate and are based on sound educational practices that consider how children learn best.

Young children are "me" learners. They are most interested in things that are relevant to their lives. They are fascinated by what is happening to "**me**", right now. **Creating Life-Long Math and Science Learners** starts with the child's immediate surroundings and branches out. The ideas are designed to make learning **meaningful**, **engaging**, **spontaneous**, and **fun** in an environment that is friendly to risk-taking. Many parents want to spoon-feed math and science concepts to children, however, many of these concepts are abstract and often above the child's understanding. **Creating Life-long Math and Science Learners** encourages parents to use **real-life** experiences and explorations to help their child make concrete discoveries. Also provided are suggestions for motivating children to becoming independent learners.

Many parents mistakenly believe that the best way to prepare their child for math and science is by completing workbooks. However, workbooks are not the answer for young children. These books do not give positive feedback, and in many cases the child receives negative reinforcement through incorrect responses. Workbooks are solitary projects and do very little to foster creativity. Workbooks are what are often referred to as "canned curriculum," with a "one-skill level fits all" approach. Young children become passive participants as they sit quietly filling in blanks, circling pictures, or drawing lines to matching objects. Throw away those workbooks, turn off the factory-generated programmed computer games, and become an active participant in your child's math and science learning experiences! Start thinking of your child as an **individual learner** with his/her own **interests** and **needs**!

The United States has been a leader in technology and has always been able to produce a workforce that is math and science literate. However, in recent years, the math and science skills of American children have steadily declined. The skill levels of American children have fallen behind many of the industrialized nations of the world, thus jeopardizing America's ability to compete on the global front. In order to prevent the furtherance of this decline, a nationwide task force of American science and math educators studied the problem. As a consequence, minimum national math and science standards and guidelines were established for children in grades Kindergarten-12 and states were expected to apply these standards to their math and science curriculums. It was envisioned that these changes would give new life to the nation's math and science educational programs.

However, these standards left out two important keys to success. **First**, waiting until a child has reached school age to focus on math and science standards is far too late. Developing an interest in math and science must begin before the child enters school. Nonetheless, school systems continue to emphasize math and science skills in the higher grades. If America ever hopes to raise its science and math levels, it must first begin to **validate** the importance of strong math and science early childhood curriculums. Somehow a fundamental law of physics seems to have been forgotten, which is that "**a building is only as strong as its foundation.**"

Secondly, these standards failed to acknowledge the vital role of **parents** in the lives and learning of children ages 2-6. Educating the young child is a partnership that begins in the home. Furthermore, **a parent is the child's first and most important teacher**! It is up to parents to make their child's first experiences with math and science exciting and meaningful. A child's excitement for learning is in many ways directly related to their parents' attitudes. If parents are engaged in learning, the child will be as well. The child's future is in the parents' hands. Therefore, during these formative years, parents must take every opportunity to optimize children's chances for success.

Across the country, American children are taught what could be called "**trivia science and math.**" They memorize just enough facts to pass a test, but do not learn enough substance to understand the total process. The emphasis is on **quantity** not **quality**. Curriculums for each grade level are loaded with mandated science and math skills that must be covered in a single school year. Therefore, young children are inundated with mountains of materials that must be learned. There is so little time, and so much to be covered. As a result, the young child's daily classroom activities consist of reading math and science from textbooks, engaging in drills, reciting math facts, memorizing isolated facts, and performing seemly magical science experiments. The child is forced to engage in **abstract** thinking and memorization of science and math concepts long before a strong foundation of **concrete** understanding has been established.

Most of the material presented to the young child is based on abstract thinking. At an early age, science textbooks introduce young children to **atoms**. Children are taught that **everything** in their world is made-up of atoms, and that atoms are so small that a special microscope is needed see them. **Abstractly**, the child is asked to mentally visualize everything being made up of atoms. The child must picture these tiny atoms while also imagining electrons, neutrons, and protons traveling around them. Just think of the frustration a young child must be experience trying to imagine that everything around him/her is made-up of these tiny atoms. Quietly, the child must be saying, "No matter how hard I try, I can't see them, yet science says they are there." Adding further to this confusion, the textbook states that when two things are rubbed together, the atoms inside of each will begin to move around, bumping into one another and giving off heat. "If I can't see them, how do I know they are bumping into each other?" the child wonders. As the child goes through the grades, he/she is asked to build other abstract understanding on top of this fragile foundation. Upon reaching the upper grades the child has acquired hundreds of abstract math and science concepts, but there is very little understanding. It is this lack of understanding that turns many children's interest off in math and science.

Learning and understanding takes place when children engage in activities that help them to **formulate a frame of reference** and **develop prior knowledge** about a topic. Children and adults use their prior knowledge and frame of reference bases as bridges to explain, understand, and to make connections to new discoveries. The goal of **Creating Life-Long Math and Science Learners** is to motivate and excite the young child to become an independent learner while having fun building strong, **concrete** math and science foundations that will eventually support abstract concepts.

All activities in **Creating Life-Long Math and Science Learners** are easy and open-ended while supporting the following learning objectives:

- To explore math and science within the **child's own world**
- To provide **age-appropriate discoveries and experiences**
- To provide a **positive learning environment**
- To provide **hands-on discoveries**
- To develop **observational skills**
- To build related **vocabulary**
- To **formulate strategies** for gathering and organizing information
- To **evaluate information and draw conclusions**
- To develop **problem-solving skills**
- To develop **critical thinking** and **analytical skills**
- To build **prior knowledge and frame of reference bases**
- To **revisit prior skills and connect them to new skills**
- To understand that **science and math are not magic**

- To **seek out** and to **extend discoveries**
- To **understand the process**

Creating Life-Long Math and Science Learners begins each new topic with a **Parent Background** section. This is a quick refresher of basic facts intended for parents to reinforce their "**prior knowledge.**" A rationale behind how children learn specific skills is also provided in this section. See the following example: The concept of **Number** is based on an abstract value and symbol system and the relationship between the two. It is the abstractness of number that makes the concept difficult for young children to understand. Therefore, keep in mind that the parent should help the child to understand the relationships between the **number name**, **number symbol**, **number set**, and **value**. The ultimate goal for the young child (2-6) is to understand that the number named (five) can be expressed by the number symbol (5), and the symbol 5 is equal to a set (group) of five individual objects. By understanding these relationships, the child is building a number comparison base that he/she will need for understanding higher-level abstract number skills.

A three-year-old is beginning to show some number understanding by showing three fingers to indicate his/her age. This child has learned to associate the number 3, by using 3 fingers. However, he/she may not have made the connection that each finger stands alone to represent one year-(1-to-1 correspondence). **Creating Life-Long Math and Science Learners** helps parents provide countless age-appropriate, concrete, hands-on experiences for the child by developing the skill of 1-to-1 correspondence, providing a foundation for number relationships and explaining number concepts. As the child continuously counts, touches, and manipulates objects into sets (groups), he/she will begin to make these connections.

Everyday, Americans hear that the national debt is in the trillions of dollars. Most adults can visualize a million or even a billion. However, the value of a trillion is even out of the conceptual range of most adults. This is because most adults have nothing in their experiences with which to compare a trillion. To the young child who has not developed a conceptual base for the number 5, 5 might as well be a trillion. Making a child memorize the number facts "5+5=10" will do little to add to this understanding if he/she has not developed some prior knowledge about the number 5. Before the child is ready for "5+5=10", he/she must understand that 10 things are more than 9, but less than 11. He/she must also have some understanding of number combinations that are equal to 10. Only with time and by exploring, discovering, and developing problem solving strategies, will the child begin to develop **logical** and **analytical thinking** skills about the number 5.

When selecting an activity from **Creating Life-Long Math and Science Learners**, pay close attention to the indicated appropriate age levels. If the age for a skill is noted as (4-5), don't expect a 2-year-old to be able to perform this task. Many of the activities are developmental. This means that before a child is ready for a task other variables must first be in place. For example, most 2-year-olds do not usually have the visual discrimination skills of a 4-year-old; therefore, most 2-year-olds are not ready to visually distinguish slight differences between two or three objects. Don't rush to teach skills that are beyond the child's understanding. There are so many other **fun** math activities outside of teaching a 3-year-old to memorize the abstract number facts "2+2=4", or "4+4=8". This is a skill that can wait until the child has a better understanding of numbers.

Discovering and **communicating** go hand-in-hand. Young children need verbal tools to describe their experiences. To accomplish this they must **develop a strong vocabulary**. Throughout this book, an emphasis is placed on vocabulary development. Have fun with words; however, the **only rule** is **that the words must be a part of a meaningful, age-appropriate experience** not an **isolated** drill. If a parent uses the word, eventually it will become a part of the child's vocabulary.

  Take for example the following:
1.   "Let's use the blocks to build the tower. We will **construct** a tower." "Can you **construct** a tower that is taller than this one?"

2.   "Put all of the red blocks over here. Now **group** the green blocks together." "Show me which things belong in this group."

• Always be aware of experiences to introduce and use new vocabulary.

• Example: "You are a member of a **group** of animals called **mammals**." In the back of the book you will find an extensive **math and science vocabulary list**. Become familiar with the list. It is appropriate for multiple ages, and most words can be applied to many different situations. Keep in mind that the goal is to help the child to develop a "prior **knowledge vocabulary** base" (with understanding). Throughout a lifetime, he/she will always be adding, extending, and connecting to this vocabulary base.

Creating Life-Long Math and Science Learners -Do's and Don'ts

Do's
• Engage in activities that are **age-appropriate**.
• Make the activity **fun** and a **natural part** of the child's daily life.
• Use **real life** experiences for making discoveries.
• Read **background**, **objectives**, and the **vocabulary** for each activity.
• Take **clues** from the child. If he/she seems involved in a topic, look for ways to extend it.
• **Revisit** a skill or concept by connecting old skills to new concepts and skills.

- Provide **hands-on** materials.
- Take advantage of every **teachable** moment.
- Provide **books** related to the topic.
- **Encourage** the child to make discoveries on his/her own.

Dont's
- Don't approach an activity as if it is a "**have to**". If the child doesn't seem interested, come back to it later or try another approach.
- Don't make the child engage in **drills**. Drilling takes the fun out of learning.
- Remember that time is on your side; **so take your time**. Too much too soon will not benefit the child!
- Mastery of a skill comes with time.
- **Mathematicians** and **scientists** are **not** created in a day.

Always remember the best gift that can be given to a child is the gift of **Time...**

-

To Talk!
- To Listen!
- To Learn!
- To Ponder!
- To Explore!
- To Think!
- To Try!
- To Question!
- To Make Connections!
- To Wonder!
- To Grow!
- To Be Creative!
- To Make Mistakes!
- To Correct Mistakes!
- To **Be Me!**

Observational Skills
Learning is a complex process that takes time to evolve!
Parent Background Development
    Techniques, skills and materials:

- **Observational Skills**
  How to use this skill

- **Comparing and Contrasting Skills**
Basic Comparison and Contrasting Questions
- **Information at a Glance**
Basic math and science knowledge on general science and math topics

- **Creating a Math and Science Box**
A list of suggested materials

Observational Skills

- Exploring math and science requires looking to note changes, similarities, differences, and evaluating these changes to understand and draw conclusions.
- As you and your child engage in math and science discoveries, ask **open-ended questions**. Questions should make the child-**stop**, **look**, and **think** critically and analytically, as he/she analyzes and compares information to make connections and discoveries.

Suggested Open-Ended Observation Questions-
- What do you see?
- Why do you think?
- What will happen if...?
- What do you think we should do?
- Next time, what shall we do different?
- How has it changed?
- Why do you think it changed?
- How can we change it?
- Tell me how you did that?
- What shall we do first?
- Why did you do that?
- What shall we do next?
- What were you trying to find out?
- Show me?
- How many?
- Are there more? Less? Same?
- How are they alike, different?
- Can you make them equal?
- Do you see a pattern?
- Can you tell me the sequence?

Comparing and Contrasting Skills

Comparing and **contrasting** are **observational skills.**
- These skills help the child to develop and use **problem solving** and **critical thinking techniques** to form **logical conclusions.**
- The child is asked to use his/her **prior knowledge** to note **similarities** and **differences**, to **evaluate** and to **draw conclusions.**
- To get the most from comparing and contrasting activities your child must be shown how to look and think critically at objects.

- Comparisons and contrasting is **not meant to be a guessing game**.
- Give the young child some visual clues-a picture or the actual objects. This will elevate wild guessing and gives the child a **frame of reference** on which to base ideas.

- **Animals**

Below are some of the possible ways for comparing and contrasting a **snake** to a **dog**.
- Step 1-List how **each** animal is **different**.
- Step 2-List how they are **alike**.

Comparing a Snake to a Dog

Snakes   Dogs

Step 1
How are they Different?

Snake	Dog
1. Scales on body	Covered with fur
2. No legs	4 legs
3. Reptiles	Mammals
4. Most lay eggs	Babies born alive
5. Do not care for young	Care for young
6. Hiss	Bark
7. Molt	Shed some hair
8. Swallows food whole	Tear and chew food
9. Cold-blooded	Warm-blooded
10. Hunts for its food	Most do not hunt prey
11. Hibernate	Does not hibernate
12. Does not need to eat daily	Must have food daily
13. Shy	Friendly

14. Fangs   Teeth
15. Uses its tongue to smell   Smells with nose
16. Most do not care for young   Females care for young
17. Does not play   Likes to play
18. Slitters   Run, walk
19. No ears   Have ears
20. Inject poison into its prey   Non-poisonous

(There are many differences that can be made, be **creative**.)

Step 2
How are a dog and a snake Alike?
1. Are alive
2. Are animals
3. Have lungs. (breathe air)
4. Carnivores (eat meat)
5. Have 2 eyes
6. Have heads
7. Have backbones(vertebrates)
8. Have babies
9. Eat food
10. Have a tongue
11. Will bite
12. Can be a pet
13. Hide
14. Have a good sense of smell
15. Move fast

Basic Comparison and Contrasting Questions:
✓ **Animals**
1. Is it a **mammal**, **reptile**, **amphibian**, **fish**, **insect** or a **bird**?
2. Does it have something that no other animal has?
3. Can it do something that no other animal does?
4. What color is it?
5. How does it breathe?
6. Is it big or little?
7. Does it have legs?
8. Does it have hair, scales, feathers, or fur?
9. Is it a farm, zoo, water, jungle, or a house animal?
10. Is it **cold or warm-blooded**? (Warm-blooded animals are able to produce body heat, cold-blooded cannot.)
11. Does it have eyes? How many? Where are they located?
12. Is it a **nocturnal** or a **diurnal** animal? (Active night or day)
13. Does it have teeth?
14. What does it eat?

15. How does it eat?
16. It is an **herbivore**- plant eater, **carnivore**- meat eater, **omnivore** meat and plant eater or an **insectivore**-insect eater?
17. How does it capture its food?
18. Do people eat it?
19. Do other animals eat it?
20. Does it have a **vertebrate? (Has a backbone)**
21. Is it an **invertebrate?** (no **backbone**)
22. Does it have an **exoskeleton?** (Hard outer skeleton)
23. Does it have an **endoskeleton**? (Hard on the inside, bones)

24. Does it care for its young?
25. Can it climb a tree?
26. Does it live above or below ground?
27. Does it dig a **burrow**?
28. Does it live alone or in a community?
29. Does it have ears?
30. Can you see its ears?
31. How many legs does it have?
32. How does it move?
33. Can it swim?
34. How does it **communicate**?
35. What kinds of sounds does it make?
36. Does it have **hooves, feet, paws, claws, flippers, fins, wings etc.?**
37. Does it live in a certain place on earth?
38. Does it **migrate**? (Travel to find food, example- birds and whales)
39. Does it **hibernate**? (Sleep or rest to conserve energy)
40. Is it **poisonous** or **non-poisonous?**
41. Does it molt? (Shed its skin-example-snakes)
42. Who are its enemies?
43. Is it friendly to people?

How does it use **camouflage?** (Skin coloring to make it difficult to see)

✓ **Plants**
1. How does it look?
2. What color is it?
3. Where does it grow? (Ground in water or on a tree)
4. Is it found only in a certain place?
5. Does it have a flower?
6. Do people eat it?
7. What part of the plant is it?
8. Is it a **seed**?
9. Is it a **stem**?
10. Is it a **root?**

11. Is it a **leaf**?
12. Is it a **nut?**
13. Is it a **fruit?**
14. Does it have **chlorophyll**?
15. Does it make its own food?
16. If it does not produce its own food, how does it feed?
17. How does it feel? (ridges, smooth, rough, bumpy)
18. Is it hard or soft?
19. Does it grow tall or short?
20. Does it offer animals **camouflage?**
21. Is it **nonpoisonous?**
22. Is it **poisonous**?
23. Is it a **coniferous** plant? (A plant that does not loose all of it leaves, makes cones)
24. Is it a **deciduous** plant?(A plant that looses its leaves in the fall)
25. Does the plant have a way of protecting itself? (Thorns)
26. Does it smell?
27. How do the seeds travel? Hitchhikers, wings, wind, animals)
28. If it is a seed, where did it grow? (fruit, pod, inside a nut shell, on a flower)
29. What animals depend upon it for survival?

✓ **Non-living Things**
1. What is it?
2. How does it look?
3. How does it feel?
4. What is it made of?
5. What color is it?
6. How is it used?
7. Can you sit on or get into it?
8. Who uses it?
9. Why is it used?
10. Where is it used?
11. Where can it be found?
12. What shape is it?
13. It is big or little?
14. Is it hard or soft?
15. Is it used inside or outside?
16. Can you smell it?
17. Can you taste it?
18. Can you wear it?
19. Can you hold it?
20. Can it hold anything?
21. What must you have/do in order to use it?
22. Does it have moving parts? What part moves?
23. Does it make a noise?

24. Is it a tool?
25. What work does it do?
26. Does it use electricity? Oil? Gas? Batteries?
27. Can it hurt you?

Information at a Glance for Parents

Below are basic science and math facts for **parents**. Keep this section in mind; it will come in handy as a quick reference of basic facts.

- **All Living Things**
  - ➢ Breathe
  - ➢ Reproduce
  - ➢ Must eat
  - ➢ Must have water
  - ➢ Grow

- **.Cold-Blooded Animals**
  - ➢ Bodies not able to maintain a steady heat level
  - ➢ Body core temperature is same as surrounding air
  - ➢ Many bask in the sun to keep warm
  - ➢ Fish, reptiles, amphibians, invertebrates (insects, spiders, worms, shellfish etc.)

- **Warm-Blooded Animals**
  - ➢ Bodies able to produce heat to keep themselves warm
  - ➢ Inside body temperature is at a constant level
  - ➢ Mammals, birds

- **Mammals**
  - ➢ Humans, horses, apes, whales, dogs, and cats, etc
  - ➢ Have fur or hair
  - ➢ Warm-blooded
  - ➢ Nurse young
  - ➢ Lungs for breathing
  - ➢ Four limbs
  - ➢ Give birth to live babies. (Except a few)
  - ➢ Have a backbone- vertebrates (endoskeletons)
  - ➢ Moving lower jaws
  - ➢ Bats are the only flying mammals.
  - ➢ Some migrate
  - ➢ Some hibernate

- **Birds**
  - ➢ Examples-Robins, ostriches, penguins
  - ➢ Have feathers

- ➢ Warm-blooded
- ➢ Have beaks- no teeth
- ➢ Lay eggs
- ➢ Most fly
- ➢ Lungs for breathing
- ➢ Most sing
- ➢ Have a backbone-vertebrates (endoskeletons)
- ➢ Some migrate

- **Reptiles**
- ➢ Snakes, lizards, turtles, crocodiles and alligators
- ➢ Cold-blooded
- ➢ Have backbones- vertebrates (endoskeletons)
- ➢ Most have scaly skin
- ➢ Most lay eggs
- ➢ Most live on land
- ➢ Breathe through lungs

- **Amphibians**
- ➢ Frogs, newts, salamanders, toads
- ➢ Smooth skin
- ➢ Have a backbone- vertebrates (endoskeletons)
- ➢ Begin life in the water/gills for breathing
- ➢ Develop legs and lungs for land
- ➢ Cold-blooded
- ➢ Lay eggs
- ➢ Begin life in water-most leave water to live on land

- **Fish**
- ➢ Examples-Fish- sharks
- ➢ Live in water
- ➢ Body covered with scales
- ➢ Have backbones- vertebrates (endoskeletons)
- ➢ Fins for movement
- ➢ Most lay eggs
- ➢ Some give birth to live
- ➢ Cold-blooded
- ➢ Breathe through gills

- **Invertebrates** (no backbone)
- ➢ Examples-Insects, spiders, worms, shellfish, clams, octopuses, worms, slugs, spiders, mollusk, snails, arthropods etc.
- ➢ No backbone

- ➢ Some have exoskeletons (hard on the outside)-shellfish
- ➢ Cold-blooded
- ➢ Some have wings
- ➢ Some fly
- ➢ Some live in water
- ➢ Reproduce in a variety of ways

- **Insects**
- ➢ Examples-Butterflies, beetles, grasshoppers
- ➢ 3 body parts- 1-head, 2-thorax, 3-abdomen
- ➢ 6 legs- Attached to thorax
- ➢ Jointed legs
- ➢ Hard exoskeleton
- ➢ Cold blooded
- ➢ Lay eggs
- ➢ Have antenna
- ➢ Most have wings-Attached to wings

- **Arachnids**
- ➢ Examples-Spiders, scorpions, ticks, mites
- ➢ 2 Body parts- 1-head and thorax fused together, 2-abdomen
- ➢ 8 legs
- ➢ Jointed legs
- ➢ Hard exoskeleton
- ➢ Most are carnivores
- ➢ Cold-blooded
- ➢ Many spin webs

- **Types of Eaters**
- ➢ Herbivores- plant eaters
- ➢ Carnivores- meat eaters
- ➢ Omnivores- meat and plant eaters
- ➢ Insectivores- insect eaters
- ➢ Parasites-live off of others

- **9 Planets**
(The Sun)
1. Mercury
2. Venus
3. Earth
4. Mars
5. Jupiter
6. Saturn
7. Uranus
8. Neptune
9. Pluto- a dwarf planet

- **7 Continents**
1. North America
2. South America
3. Europe
4. Asia
5. Africa
6. Australia
7. Antarctica

- **Polar Regions**
➤ Antarctica- South Pole
➤ Arctic- North Pole

- **Dry and Wet Measurements**
➤ 16 Tablespoon =1 cup
➤ 8 ounces = 1 cup
➤ 2 cups = 1 pint
➤ 16 ounces = 1 pint
➤ 2 pints = 1 quart
➤ 32 ounces = 1 quart
➤ 4 quarts = 1 gallon

- **Linear Measurements**
➤ 12 inches =1 foot
➤ 3 feet = 1 yard
➤ 36 inches = 1 yard
➤ 1760 yards = 1 mile
➤ 5,280 feet = 1 mile

- **Weight Measurements**
➤ 16 ounces = 1 pound
➤ 2000 pounds = 1 ton

- **Time Measurements**
➤ 60 seconds = 1 minute
➤ 60 minutes = 1 hour
➤ 24 hours = 1 day
➤ 7 days = 1 week
➤ 30 or 31 days = 1 month
➤ 12 months = 1 year
➤ 365 days =1 year–leap year 366 days every four years
➤ 10 years = 1 decade
➤ 100 years = 1 century
➤ 1000 year = 1 millennium

- **Temperature Measurements**
  - Human's normal body temp. -98.6 Fahrenheit, 37 degrees Celsius.
  - Water freezes at 32 degrees Fahrenheit, 0 degrees Celsius.
  - Water boils at 212 degrees Fahrenheit, 100 degree Celsius.

- **The Four Seasons**
  - Summer- June 21st–September 20th
  - Fall- September 21st–December 20th
  - Winter- December 21st–March 20th
  - Spring- March 21st –June 20th
  - 30 days in September, April, June. November
  - 31 days in January, March, May, July, August, October
  - 28 days February (29 days leap year every four years)

Creating a Math and Science Box

Creating this box will mean that materials will be available when needed. It is "always a work in progress." The materials will change as new concepts and skills are added. You and your child decide what is needed. Prepare a smaller version for the car.

Include:

- Paper of various sizes and colors
- A set of magnetic numbers and shapes
- Crayons
- Pencils
- Colored Markers-broad and fine tip
- Tape
- Rulers
- Yard stick
- Tape measure
- Blank index cards
- Glue
- Scissors
- Playdough
- Package of Poker chips
- A set of number cards 0-20
- Shapes cut from cardboard
- Magnifying glass
- Binoculars
- Magnets
- 2 and 3 column graphing mats
- Paper clips
- Books identifying trees, insects, birds, flowers, mammals, the universe, etc.
- A small clear jar, netting, and rubber bands (to cover the jar) for collecting and observing small specimens

Mary Taylor Overton

- A small shovel
- Flashlight

Left and Right

Mary Taylor Overton

Right and Left

Left   Right

Parent Background

- Right and left are skills used to identify **position** and **direction**.
- For all of the following activities, provide a frame of reference. Example: 1. Mark the right side of the room with a large star. Begin each **activity** by facing the star.
- When identifying right or left, pay close attention to your position. When you are facing your child, your right is his/her left. You must raise **your left** hand to indicate his/her **right**.
- Begin all activities by making sure that you establish **right** from the **left**. Be **consistent**, always begin with a clue.

Suggested Right and Left Activities- (2-6)

- Use 2 different color socks to make **right** and **left**-handed puppets. Add button eyes and decorate. Use the puppets along with the suggested activities.
- Think of other fun things to do with the puppets.

Right-Side Week -For a week emphasize **Right**. ("Left Side Week," should come after your child has learned to identify his/her right) (3½-6)
- 1. Put a washable tattoo on your child's **right** hand.
- 2. Trace your child's **right** foot and hand. Compare the left and right foot and hand. Note the position of the thumbs, large toes and other fingers and toes.

- 3. Put an X inside your child's **right** shoe. .

- 4. Use your right hand to pick up the toys on the **right** side of the room.

- 5. Put your **right** sock on last. Walk about the house wearing the right sock and shoe.

- 6. Put your **right** shoe on first.

- 7. Sit on the **right** side of the car.

- 8. Place the napkin on the **right** side of the plate.

- 9. Put your fork on the **right** side of the plate.

- 10. Put your child's food on the **righ**t side of the plate.

- 11. Dip your **right** hand in washable paint to make handprints on white or colored paper. Use the paper to wrap a gift. Try your right foot.

- 12. Put on your **right** mitten (a clean sock may be substituted) and try doing fun activities.

- 13. Fold a large sheet of paper in $\frac{1}{2}$; draw a line down the centerfold. Draw a picture on the **right** side of the paper.

- 14. Get into bed from the **right** side.

- 15. Brush the right side of **your** teeth first.

- 16. Touch things that are on the **right** side of the room.

- 17. Put all of the magnets on the **right** side of the refrigerator. Take all of the magnets off of the right side.
- 18. Turn the pages of a book with your **right** hand.

- 19. Use your **right** hand to identify everything that is on the **right** side of your body. Start at your head and move down to your toes.

- 20. Try counting things using your **right** hand.

- 21. Wear a toy ring on your **right** hand.

- 22. String beads to make bracelet and wear it on your **right** wrist.

- 23. Make a large nametag to wear on the **right** side of your chest.

- 24. Wear a toy watch on your **right** wrist.

- 25. Make a washcloth mitt from a lightweight wash cloth. Fold the cloth in half and sew the bottom and side seams. When you take a bath use the mitt on your **right** hand.

- 26. As you walk, balance something in your **right** hand.

- 27. Divide a sandwich in half and eat the **right** half-last.

- 28. Order a pizza and eat the **right** half-first.

- 29. Walk on the **right** side of the street.

- 30. Roll play dough with your **right** hand.

- 31. Bake cookies and put icing the **right** side and decorate with raisins, nuts, candy, or chocolate chips.

Geometry

Geometry

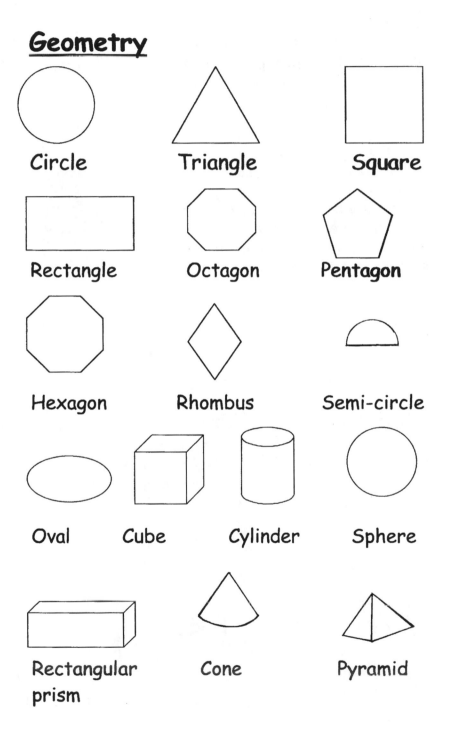

Circle　　　Triangle　　　Square

Rectangle　　　Octagon　　　Pentagon

Hexagon　　　Rhombus　　　Semi-circle

Oval　　Cube　　Cylinder　　Sphere

Rectangular prism　　Cone　　Pyramid

Geometric Shapes-(2-6)

Parent Background-(2-6)
- **Geometry** can be defined as relationships between the properties of **lines**, **symmetry**, **angles**, **surfaces**, and **space**.
- Lines intersect and connect to form many different **shapes**.
- Geometric shapes are an important part of your child's world. Throughout life he/she will come into contact with geometric forms.

Lines (2-6)
Parent Background

- Everything in our world that we can see is made-up of lines.
- Children as young as 2 can have fun exploring and discovering lines and their properties.
- All of the following activities are open-ended. They are also excellent for developing fine motor skills (strengthening finger muscles).
- Most 2 - 3-year-old children do not have the fine motor skills needed for holding a pencil or a crayon; encourage the use of a finger instead.
- ✓ Make lines in the air.
- ✓ Put shaving cream on a table and make lines.
- ✓ Make lines in sand, rice or mud.

Vocabulary Development-

- Lines
- Straight lines
- Zigzag lines
- Diagonal lines
- Long line
- Short line
- Fat / thick lines
- Skinny/ thin lines

- Vertical lines
- Horizontal lines
- Wavy lines
- Crossing lines
- Dotted lines
- Connected lines
- Broken lines

Suggested Line Activities (2-6)

1. Use your finger to make lines in the air. As you make the lines, your child is to copy. Take turns being the leader. "I am making a **zigzag** line." (2-6)

2. Many homes have **vertical** blinds, if you are so lucky; discuss why they are called, vertical blinds. Refer to regular mini-blinds as **horizontal**. Compare the

two. How are they alike? Different? (2-6)

3. Curtains hang **vertically**. Use your finger to trace the vertical lines of curtains. Find things around the house that are vertical- (Wallpaper, doors, walls, windows, cabinets etc.). (3-6)

Vertical Things
- People stand vertically
- Telephone poles
- Car antennas
- Street lights
- Mirrors
- Buildings

Horizontal Things
- Boats
- Rugs
- Beds
- Airplanes

- 4. Go for a walk in the neighborhood to look for things that are **horizontal** or **vertical**. (3-6)

- 5. (3-6) Fold a piece of paper in half lengthwise. Open and fold again widthwise, so that you have 4 boxes. Let your child use a marker to trace the **vertical** fold line. Use a different color to trace the **horizontal** line. Make pictures in each of the boxes.

- 6. Use **fine** and **broad** tip markers to make **thin** and **fat** lines. Use the markers to make pictures. (2-6)

- 7. Make lines on the sidewalk with sidewalk chalk. (2-6)

- 8. Cooked spaghetti is excellent for exploring different kinds of lines and shapes. (short, wavy, long, crossing etc.) (2-6)

- 9. Use your finger to make lines in the air. Make crisscrossed (X), large, small, backwards, upside down, and sideway lines. (2-6)

- 10. Paper clips hooked together can be used to make long, short, curved, and wiggly lines. (2-6)

- 11. Fill a bucket with water and provide paintbrushes of different sizes to have fun making lines on outside walls or sidewalks. (2-6)

- 12. Make **curved** lines and continue the lines into circles or other shapes. (2-6)

- 14. Use a ruler to make **straight** lines. (3-6)

- 15. Make your own dot-to-dot pictures. (3-6)

- 16. (4-6) Study the **constellations** in the sky. Learn the names of some of these constellations.

- 17. Walk in a **straight** line from one point to another. Try walking and making zigzag, curved, or wavy lines. (3-6)

- 18. Use dried beans or small cereal to make wavy, straight, curved, long, short, wide, or thin lines. (3-6)

- 19. Pieces of string can be used to create different kinds of lines. (3-6)

- 20. During bath time, use **washable markers** to trace the vertical and horizontal lines on the tiled wall. (2-6)

Symmetry (3-6)
Parent Background
Most geometrical shapes are **symmetrical**. This means that the shape is equal on corresponding sides. Your child will have many opportunities to refine his/her symmetry skills.

Vocabulary-
- Symmetrical
- Equal
- Same

Suggested Symmetry Activities (3-6)

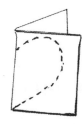

Heart

- 1. To make a symmetrical circle, fold a piece of paper in $\frac{1}{2}$. Draw a semi circle on the fold. Keep the paper folded and cut out the semi-circle. Unfold to reveal a symmetrical circle.

- 2. Make symmetrical hearts.

- 3. Prepare a basket of pre-folded paper, crayons, and a pair of scissors. Let your child make his/her own free-form symmetrical shapes. Show how to draw a shape and cut. Give your child time to explore and make discoveries on his/her own.

- 4. Play a game using blocks or other shapes. First lay 1 (red) center block on the table; this will be the starting point (the center). Add two different color blocks one on each side of the center block. Your child is to add 1 more blocks to each side. Continue taking turns, always adding blocks to keep the shape symmetrical.

- 5. Find things around the house that are symmetrical.

- 6. Make a symmetrical necklace. On a long shoestring, add a red center bead. Next add a blue bead to each side of the red bead. Continue adding corresponding colored beads to each side of the string.

- 7. Help your child to discover that his/her body is symmetrical.

- 8. **Patterns for Making a Symmetrical Person.**

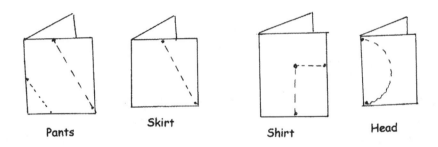

Pants   Skirt   Shirt   Head

Materials- Construction paper-1 flesh tone colors for the face and 2 different colors (1-for pants or skirt, 1-for shirt) glue and scissors. Fold paper in half. Follow the pattern above for the clothing. Draw and cut on the lines. Put the pieces together to make a boy or girl.

Shapes

Parent Background (2-6)
- Identifying basic shapes is generally the child first introduction to **geometry**.
- Shapes can be divided into two categories-
 2 dimensional (**flat shapes**) and 3 dimensional (**solid shapes**).

- Examples at the beginning of unit-

 2 Dimensional Shapes

Mary Taylor Overton

- -Circle
- -Triangle
- -Rectangle
- -Square
- -Hexagon

- -Semi-circle or half circle
- -Pentagon
- -Octagon
- -Oval
- -Rhombus

Solid Shapes or 3 dimensional shapes-
- -Spheres
- -Cubes
- -Cylinders
- -Rectangular prism
- -Cones
- -Pyramids

As your child progresses through the grades he/she will come into contact with variations of these shapes.

Before beginning geometry prepare a shape collection-
- -Cut shapes from a variety of textured paper.
- -Put some in your **math and science box**. Also prepare a set for your car, this way they will always be available. Just pull out a shape for an easy reference.
- -Cut shapes from colored Contact paper and adhere to a wall in your child's room.
- -Make shape place mats. Glue shapes on a large piece of construction paper. Laminate or cover with clear contact paper. During mealtime play a game, "Can you find a triangle?"

Circles–(2-6)

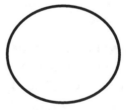

in

Circles are the easiest of the geometric shapes to identify. They can be found abundance throughout your child's world.

Objectives-
1. To identify the properties of **circles**
2. To make circles
3. To discover circles in the environment

Vocabulary
- Circle
- Shape
- Circular shape
- Circular motion
- Round
- Curved line
- Corners

Parent Background (2-6)

Developing Circle Vocabulary-

- With a little help from an adult, children as young as 2 can draw circles.
- Take your child's finger and help him/her to make a circle in the air, while saying, "This is a circle."
- Your continuous use of the vocabulary will help your child to make it a part of his/her own descriptive language.
- Drawing a circle in the palm of his/her hand will encourage him/her to try reproducing the same shape, on the child's own. " Watch me make a curved line. Can you make a curved line into a circle?"
- (3½-6) Help your child to describe the circle (No sides, no corners, it is round like a tire, sun, plate)
- It is important to keep **talking** about circles.
- **Use the vocabulary as much as possible. Take your time.**
- Your child does not have to master the skill today or tomorrow! In time and with lots of experiences, he/she will be able to identify the properties of different shapes. (2-6 year old)

Going On A Circle Hunt-You are now ready **to hunt for circles**. The kitchen is a good place to begin. From there move to other parts of the house and the neighborhood. Select circles that are safe for small fingers.

Plates	Cut Hot Dogs	Moon
Pancakes	Measuring Cups	Sun
Buttons	Pots Lids	Steering Wheels
Clocks	CDs	Watches
Doorknobs	Top of Trash Cans	Nail Heads
Unopened Cans	Coffee Tables	Buttons
Watermelon Slices	Patio Tables	Doorbells
Orange Slices	Mirrors	Wais
Bottle Caps	Rugs	Yo-Y
Plastic Containers Tops	Money	Rings
Round Baskets	Cookies	Lollip
Placemats	Clocks	Pita l
Potato Slices Cucumber	Vegetable Slices Clocks	
Slices Tomato Slices	Sunny Side Eggs	
Tires	Cooked Shrimp	
Wheels		

Suggested Circle Activities- (2-6)

- 1. Circles cut from sponges are excellent for use in the bathtub or pool and they float. (2-6)

- 2. **Use body parts to make circles**- hands, fingers, legs, feet, head, etc. (2-6) "Can you think of another way to make a circle?" " How can the 2 of us

29

make a circle together?" " Can you think of a way that we can make a larger circle?"

- 3. **Have a Circle Dinner.** (2-6) Everything served must be in the shape of a flat circle, not round like a ball. Make circle place mats and coasters. **Suggested menus**-Flattened meatballs, hot dog slice, sliced tomatoes, cucumbers, carrot slices, potato slices, onion rings, fruit slices, cupcakes, cakes, or cookies. Before eating have fun talking about the items being served.

- 4. Join hands and **play circle games.** "Ring-a-Round the Rosies", "The Farmer in the Dell." ($1\frac{1}{2}$-6)

Ring-a-Round the Rosies

  Ring-a-Round the Rosies
  Pocket full of Posies,
   Ashes, ashes,
  We all fall down.

- 5. Make sandwiches on **pita bread.** ($1\frac{1}{2}$-6)

- 6. Sort and string **circular buttons.** (2-6)

- 7. Eat round **rice cakes.** ($1\frac{1}{2}$-6)

- 8. Serve **mini bagels** and **donuts** ($1\frac{1}{2}$-6)

- 9. Make a fruit salad using **round fruit slices.** Select such fruits as oranges, grapefruits, bananas, kiwi, and guava. ($1\frac{1}{2}$-6)

- 10. Bake **cookies, cakes, pies,** and **cupcakes.** ($1\frac{1}{2}$-6)

- 11. Make hamburgers and serve on **round hamburger rolls.** Add round cheese slices and pickle rounds. Let your child help to make the hamburgers. "How, can we make the hamburgers into a circle?" ($1\frac{1}{2}$-6)

- 12. Eat snacks on **circular crackers.** (2-6)

- 13. **Freeze circle** ice treats. Ask your child to select a suitable **circular** container. A muffin tin is excellent for this project. Add a small amount of juice or chocolate milk and freeze over night. Plan ahead. (2-6)

- 14. Make sandwiches from **circular cold cuts** and **cheeses.** (2-6)

- 15. Cut **sausages** into circles. (2-6)

- 16. Sort soda or water bottle **plastic tops**, by color, size (2½-6). Can you find all of the blue tops?

- 17. Serve foods made on round **tortillas**. (2-6)

- 18. Round **English muffins** make excellent individual pizzas, cheese toast, or sandwiches. (2-6)

- 19. Cut **vegetables** into circular slices- carrots, cucumbers, and beets. (2½-6)

- 20. Make round **potato pancakes**. (2-6)

- 21. Enjoy a **circle lollipop**. (2-6)

- 22. Bake canned prepared **biscuits**. (2-6)

- 23. Sort and count a small bag of **assorted colored candy**. Sort by color and then count them.

- 24. Make small silver dollar sized circular **pancakes**. (2-6)

- 25. Make and decorate a crown to wear. "How can you make this strip of paper into a **circular crown**?" (2-6)
- 26. Walk, run, hop, crawl, and skip in circles. "Can you think of other activities that we can do as we **move in a circle**?" (2-6)

- 27. Provide child-sized paintbrushes and a bucket of water to make circles on the sidewalk or a brick wall. (3-6)

- 28. Use sidewalk chalk to make circles on the sidewalk. It will easily wash away with a water hose or rain. (3-6)

- 29. Wash the table in a **circular motion**. (2-6)

- 30. Decorate paper plates to use as **flying saucers**. (2½-6)

- 31. Cut ½ inch wide paper strips to make a **connecting chain of circles**. "How can we connect them to make a long chain?" (3½-6)

- 32. Cut circles of different sizes and use to make pictures. Examples- snowman, a clown's face, etc. "What can we make from these circles?" (2-6)

- 33. Use paper plates to make **animal** and **people face masks**. (2-6)

- 34. Investigate the **symmetry of circles**. Fold paper in half. Draw a semi-circle on the folded line and cut along the line. Open and you have a symmetrical circle. (ages 4-6 can cut alone; $2\frac{1}{2}$ should watch as you cut)

- 35. Roll plastic circular **plates** or **deli container lids etc.** Try a variety of surfaces- smooth, bumpy, up hill, down hill, anywhere. (3-6)

- 36. Dip the circular tips of plastic hair rollers into food coloring or colored tempera paint. Stamp the colors on white paper to make designs. Try using medicine bottles, cookie cutters, glasses etc. (2-6) and patterns.

- 38. While taking a bath make circles on the tile using washable markers or colored chalk. ($2\frac{1}{2}$-6)

- 39. **Find things that turn in a circular motion**-bottle caps, fans, planets, wind-up toys, watch hands, blinders, hot and cold water facets, tires, records, CDs, clocks, bicycle wheels, windmills, Ferris wheels, globes, pinwheels, steering wheels, light bulbs, egg beaters, mixers, rolling pins, door knobs, merry-go-rounds, spinning tops, helicopters, roller blades, lighthouse lights, etc. (2-6)

- 40. Cook things that require **stirring in a circular motion**-
✓ Bake a cake-mixing cake batter
✓ Hot chocolate-stirring hot chocolate
✓ Making punch-stirring to mix
✓ Pancakes-stirring to mix pancake batter
✓ Anything that requires stirring

- 41. Make a "**circle cap.**" Cut circles from felt. Use at least 2 or 3 different colors. Use hot glue or sew the circles to attach to an old baseball cap. Let your child decide on the design. Encourage him/her to wear it. Make one for other members of the family. (2-6) Create a matching circle shirt; by attaching felt circles to an old shirt.

- 42. Observe and draw the **sun** and the **moon**. (2-6)

- 43. Use **self-sticking colored dots** to decorate a project. These dots are found in the stationary section of stores. (2-6)

- 44. Play **ring toss**. (2-6)

- 45. Buy a **hula-hoop**. Young children can have fun rolling, running around, sitting in, or operating the hula-hoop. They can also use it as a target for a beanbag toss. "Can you think of other uses?" (2-6)

- 46. Make a **paper plate clock** with hands that move. (4-6)

- 47. Before dinner, have a **Circle Hunt**. Tell each family member to look around the house to find a circle. They must show their circle to be admitted to the table. (3½-6)

- 48. **Make circles using cooked spaghetti.** (3-6) "Can you make this piece of spaghetti into a circle?" "Can you make a smaller one?" "How can you make an even bigger circle?" Try using 2 or more pieces of spaghetti?

- 49. Decorate crafts with **circular sequins**. (3-6)

- 50. Make **circular pillows or beanbags.** Cut 2 circles from solid colored fabric. Decorate using fabric paint. Sew around the edge, leaving a small opening to stuff with cotton or beans. Sew opening. Use the beanbags for games. (2-6)

- 51. **Hide circles** around the house and go on a hunt. Make clues to lead your child to the circles. Hide a special treat at the last circle. (2-6)

- 52. To make circles, trace different sized bowls, plates, and glasses etc. (3½-6)

Semi-Circle-Half-Circle-(3-6)
Vocabulary Development
- Semi-circle
- Half of a circle
- Straight line
- Curved

Parent Background (2-6)
Developing Vocabulary-
Cut a circle in ½ to make 2 semi-circles. One side of the semi-circle is a curved line the other side is a straight line.
Most of the circle activities may also be used for semi-circles.

Suggested Semi-circle Activities-(2-6)
- 1. **Serve semi-circles for dinner.** Make semi-circle paper mats and name place cards. Be creative with the menu. Cut hamburgers, hot dogs, potato slices, cucumbers, tomato slices, carrot slices, cup cakes, and cookies into semi-circles. (3-6)

Mary Taylor Overton

- 2. **Order a pizza**, half cheese and half pepperoni. (3-6)

- 3. Make semi-circle tracers by cutting circular deli lids in half. (3-6)

- 4. Cut a variety of circles of different sizes and colors. Cut the circles into semi-circles. Try making whole circles by putting 2 semi-circles together again. This time use two different colors. (3½-6)

- 5. Draw large circles. Divide the circles in half; decorate each ½ with a different design. (3-6)

- 6. Slice a tomato in half and broil it for dinner. (3-6)

- 7. Before serving pancakes cut in half. (4-6 may do the cutting)

- 8. Cut round fruit into slices and cut in ½. (2-6)

- 9. Cut cookies in ½. (2-6)

- 10. Bake a single 8 or 9-inch layer cake. Cut the layer in half and frost one layer. Add the other half and frost the entire cake. Enjoy your semi-circle of a layer cake. Many stores sell halves of a cake. (3-6)

- 11. Cut a watermelon into slices and then in half. (3-6)

- 12. Make semi-circle plates by cutting round plates in half. Serve circle sandwiches that have been cut in half. (2-6)

- 13. Serve favorite foods in half of an empty orange shell. Cut a whole orange in half. Remove the fruit segments. Fill the orange half with fruits, meat cubes, Jell-O, puddings etc. (3-6) "How can we make this whole orange into a semi-circle?"

- 14. Trace a semi-circle on to a large sheet of paper. Use the shape to create an object-Example- a turtle, add a head and legs. -A boat or an umbrella. (3-6)

- 15. Use circular cookie cutters to make circular sandwiches on 2 different kinds of breads. Use whole wheat for the top and white bread for the bottom. Cut

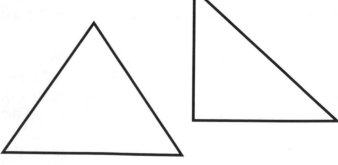

the sandwiches in half. (3-6)

Triangles (3-6)
Vocabulary Development
- Triangle
- Angles or corners
- Triangular shape
- Lines
- Sides
- Straight Lines
- Points

Parent Background
(2-3½) Should recognize triangles by sight. (3½-6) Call attention to the attributes of triangles- (**3 corners-angles** and **3 straight sides**).
Use the same introduction for **triangles** as presented in the circle section.

Place 2 straws together to make an **angle**.
- "Let's add another side to make a shape called a **triangle**. Can you make a triangle?" A triangle has 3 straight lines and 3 angles.

Going On a Triangle Hunt-
Find-

• Nacho Chips	• Arrow Heads
• Slice of Pie	• Traffic Signs
• A Slice of Cake	• Traffic Cones
• Teepee	• Birthday Hats
• Tent	• Bugle Corn Chips
• Outside Attic Roof	• Evergreen Trees
• Triangular Shaped Crackers	• Roof of a House

- Each Section of an Umbrella (Rain and Patio),
- Upside-down Capital Letter A, Letter V – W- Y- X

Suggested Triangle Activities-(2-6)
- 1. Make a square pizza. "How can we cut the slices into triangles? " (2-6)

- 2. Bake a square cake and cut into triangles. (2-6)

- 3. Serve sandwiches that have been cut into small triangular pieces. (2-6)

- 4. Cut round fruit slices into triangles. (2-6)

- 5. Enjoy a **Triangle Shape Dinner.** Serve items that have been cut into triangular shapes. Make triangular place mats. Suggestions- triangle shaped hamburgers, sandwiches, bread, green beans shaped in a triangular form. (2-6)

- 6. Cut square cold cuts and cheese slices into triangles. (2-6)

- 7. Fold napkins into triangles? (4-6)

- 8. Make a triangular shaped puzzle. Cut a large triangle from poster board. Let your child make a picture on the triangle and cut into puzzle pieces. Before cutting determine your child's level. Do not cut the puzzle into too many small pieces. Have fun putting the puzzle back together again. (3-6)

- 9. Make **teepees** and decorate. (3-6)

- 10. Draw triangles on the sidewalk. Walk on, jump in, and hop around the triangles. (4-6)

- 11. Connect 3 dots to make triangles. Have fun connecting the dots. Draw something inside. (4-6) "Let's make more."

  12. Find things in your house that can be used for constructing triangles.

- 13. Draw upside-down triangles and make them into **Christmas trees** and decorate. (4-6)

- 14. Provide paper squares of different sizes for folding triangles. "Can you can fold this paper into a triangle?" (4-6)

- 15. Cut large paper triangles and try folding them into smaller ones. (4-6)

- 16. Draw a **kite**; decorate each triangle using a different design. Attach a string and fly your kite. Purchase a commercial kite kit. As you construct the kite discuss the triangular shapes.

- 17. Glue triangles on paper and use as a part of a picture. Example-Top of a rocket, top of a tree, roof of a house, and a birthday hat. (4-6)

- 18. Make a **triangle shaped book**. Cut 5 or 6 pieces of paper into large triangles and staple into a book. (2-6)

- 19. Carve a pumpkin; make **triangular** eyes, nose. and teeth. ($2\frac{1}{2}$-6).
  Can you draw a triangle?

- 20. Create paper plate animals. Glue **small or large triangular ears**. Color or paint the plate to make desired animal. ($2\frac{1}{2}$-6) Give a puppet show.

- 21. Cut triangles of different colors and sizes. **Sort** the triangles by color or size. (3-6)

- 22. Use the triangles from activity 21 to connect triangles to make a large continuous figure. Each triangle must touch the side of another triangle. "Can you add triangles to my design?" ($2\frac{1}{2}$-6)

- 23. Write a message to someone on triangular shaped paper. (3-6)
- 24. Cut triangular **place mats** to use at mealtime. (2-6)

- 25. Make a boat, add a **triangular sail** and decorate. (3-6)

- 26. Use paper to make a **pennant**. Let your child decorate it and hang in his/her room. (2-6)

- 27. Have a snack of triangular shaped corn chips. (2-6)

Rectangles-(2-6)
Vocabulary Development
- Rectangle
- Rectangular shape
- Square
- Long side
- Short side
- Corners
- Angles

Parent Background
Vocabulary Development (2-6)
A rectangle has 4 **right** angles. It is a square that has 2 long sides and 2 short sides.
- **Refer to the introduction of a triangle and circle for other suggestions.**
- Talk about the attributes of a **rectangle**. A rectangle has 4 straight sides-

(2 long sides and 2 short sides). Rectangles and squares are so similar; children often confuse the two shapes.

- (3-6) **Compare** the attributes of **rectangles** to those of **squares**. Note that a rectangle and a square have 4 corners and 4 sides. "What do you notice about the sides?" "Are they the same?"
- Construct a rectangle from straws. Use 3 straws, cut one in ½. Make a rectangle using the four pieces. Call attention to the 2 long sides, 4 corners, angles, and 2 shorter sides. (4-6)

Going On a Rectangle Hunt-
Find-

Windows	Cereal Boxes	Sofas
Beds	Refrigerators	Bath Tubs
Mirrors	Cookie Sheets	Cards
Letters Cabinets	Sinks	Money
Tables	Cabinet Drawers	Candy Bars Credit
Papers	Computer Paper	Cards Videos
Curtains	Beds	Checks
Books Newspapers	Grocery Bags Walls	Window Shades
Magazines	Pictures	Cassettes
Rugs	Light Switch Covers	Suitcases
Towels	Envelopes Remotes	Flags
	Buses	Cameras

Suggested Rectangle Activities-(2-6)

Look through the activities for circles and triangles. Many of these activities can be applied to **Rectangles** and **Squares**.

- 1. **Construct rectangles** from straws, toothpicks, pretzel rods, sticks etc. (3½-6)

- 2. Provide a rectangular **disposable camera** for taking pictures. After developing the photos discover that they too also rectangles. (3½-6)

- 3. Put shaving cream on a rectangular shaped cookie sheet and have fun making pictures. (2-6) Before starting, use your child's finger to trace around the edges of the pan.

- 4. Have a **picnic on a towel**. Compare the 4 sides. (2-6)

- 5. Go for a ride to find buildings that are rectangular shaped. (3½-6)

- 6. **Make your street sign**. Cut rectangular cards and reproduce your own street sign. ($3\frac{1}{2}$-6)
- 7. Create a rectangular shaped book. (4-6) Staple sheets of paper together to make a book.

- 8. Help your child to make and decorate a **nameplate** for his/her bedroom door. Encourage him to make one for each family member. (3-6)

- 9. **Provide a clipboard** for drawing. (2-6)

- 10. Sort a **deck of playing cards** by color or number. (4-6)

- 11. Make pictures from rectangle shapes. Glue a rectangle on a sheet of paper. Use each rectangle as the base for a picture. Some suggestions are-
✓ -A truck- add a cab and wheels
✓ -A door- add a window and a doorknob
✓ -A wagon- add wheels and a handle
(4-6)

- 12. Make rectangular **greeting cards** for friends. Fold a piece of paper in half lengthwise into to a rectangle. Decorate the card and write a message inside. Provide a selection of different size envelopes from which to choose. "Can it fit into this one? Why not?" ($3\frac{1}{2}$-6)

- 13. Count the number of steps in your house. (2-6)

- 14. As you make your **bed**, walk around the mattress. Use 2 strings to measure each of the sides. Compare the 2 strings. (4-6)

- 15. Read **rectangular shaped books**, magazines, and newspapers. (1-6)

- 16. **Connect dots** to make rectangles. (2-6)

- 17. **Make paper money**. (3-6) Provide paper that has been cut to the size of real paper currency. Add numbers and let your child draw pictures. Provide a rectangular shaped wallet to carry the money.

- 18. Sleep on a **rectangle pillow**. ($1\frac{1}{2}$-6)

- 19. Eat **candy bars** or **fruit bars**. (2-6)

- 20. Look in the toy box to find things that are rectangular shaped. (2-6)

- 21. Make rectangular paper **flags** and have a parade. ($1\frac{1}{2}$-6)

- 22. Decorate brown **paper lunch bags** to use for carrying lunch. Make one for each member of your family. (2-6)

- 23. Eat **rectangle crackers** as treats. (2-6)

- 24. Provide a **small suitcase** for dramatic play. (2-6)

- 25. Decorate a **shoebox** and use it for a storage container. (2-6)

- 26. Make a picture on **rectangular shaped paper**. Hang it on your rectangular door. (2-6)

- 27. Remove the crust from **sandwiches** and cut into rectangular halves. (3-6)

- 28. Make **place mats**. (2-6)

- 29. Decorate a picture frame by gluing on shells, beans, pasta, etc. Add a photo and give it away as a gift. ($2\frac{1}{2}$-6)

- 30. Look at a map of your city to discover the rectangular shape of most city blocks. ($3\frac{1}{2}$-6)

- 31. Wash the table with a rectangular shaped sponge. (2-6)

- 32. Bake a rectangle sheet cake. (2-6)

- 33. **Bake brownies** and cut into rectangles. (2-6)

- 34. Decorate a legal size **file folder**. Seal edges with tape, to make an envelope. Use it for storing important things. ($3\frac{1}{2}$-6)

- 35. Build with rectangular blocks. (2-6)

- 36. **Fold napkins** into rectangles. (3-6)

- 37. Make special greeting card for a friend. (3-6)

- 38. Use a paper towel instead of a napkin. (2-6)

- 39. Cut 5-10 rectangles of different sizes. Order rectangles from small to large ($3\frac{1}{2}$-6). Reverse; order from large to small. (2-6)

- 40. Draw tall buildings. (2-6)

- 41. Examine the **windows** in your house. (2-6)

- 42. Make a **bookmark**. (3-6)

Squares (2-6)

A **square** is any shape with **4 right angles**; all of its sides are the same length.

Vocabulary Development
- Squares
- Rectangles
- Corners
- Equal or the same
- Sides

Parent Background (2-6)

Vocabulary Development
- When introducing the **square** use the same procedure as rectangles.
- Distinguishing features of the square are (4 equal straight sides 4 corners. Use 4 straws to make a square.
- Compare the rectangle's sides to those of a **square**. Use the straws from the rectangles to make the comparison.
- Make sure that your child does not confuse **squares** and **rectangles**. Keep reminding him/her that the sides of a square are **equal** or the **same**. Always provide sample square and rectangular clues. (3½-6)

• Going On A Square Hunt-

Sandwich Bread	CD Covers	nto Squc
American Cheeses Cold Cuts	Sidewalk Squares	Graham Boxes
Boxes	Napkins	Floppy D
Books	Cut Sticks of Butter i	Tile Squ Crackers

Suggested Square Activities-(2-6)

- 1. Eat square snack crackers. Provide **square**, **rectangular**, and **circular** crackers from which to choose. (2-6)

- 2. Make sandwiches from square sandwich loaf bread. (2-6)
  Cut sandwiches into smaller squares. (2-6)

- 3. Bake a **square** or **rectangular** cake and cut it into smaller squares. (2-6)

- 4. Prepare sandwiches made with **square cheese** (2-6)

- 5. Remove the square paper insert from old **CD covers**. Cut sheets of plain paper the same size and make your own covers. (3-6)

- 6. Cut several pieces of paper into large squares and staple to make a square book. Use the book as a journal. (3½-6)

- 7. Connect dots to make squares. (3½-6)

- 8. Glue a square on a piece of paper. Use the square to create pictures. Examples- (4-6)
✓ -A jack-in the-box-add a curly string and a head with a pointed hat.
✓ -A house-add a roof and doors
✓ -A book-add a picture and a title
✓ -A gift-add ribbon and a design on the box etc.
   Give your child some ideas or draw along.
- 9. Cut large squares and fold the square into smaller squares. (3½-6)

- 10. **Sort squares**. Use a variety of different colored paper to cut squares into different sizes. Sort the squares by color and size (3-6)

- 11. Count the square tiles across a kitchen or bathroom floor. There are many different size squares to discover in the bathroom. Try finding other squares. (2-6)

- 12. **Make a quilt**. Cut 12"x 12" large squares from white or any solid color fabric. Use fabric paint to draw pictures on each square. Sew squares together to make a quilt. (3-6)

- 13. Take a walk and count the blocks on your street. (3-6) square sidewalk

- 14. Cut small 1" or 2" squares different colors. Glue the make pictures. These small squares are excellent for making mosaics. "What can you make using these squares?" (4-6) from paper of squares together to

Oval-Elliptical (2-6)
Vocabulary Development
- Oval
- Oblong
- Elliptical

Parent Background-
Vocabulary Development
Ovals are elongated circles.

Going on an Oval Hunt

A Face

Almonds

Tops of Spoons

Plum Tomatoes

Olives

Footballs

Peanuts

Jellybeans

Limes

Watermelons

Lemons

Potatoes

Cameo Jewelry

Avocados

Boiled Eggs

Tables

Mangos

Pictures Picture Frames

Grapes

Spaghetti Squash

Guavas

Coconuts

Suggested Oval Activities-(2-6)
- 1. Make a **cameo**. Cut ovals, decorate and add a string. ($3\frac{1}{2}$-6)

- 2. Make a drink from **lemons** or **limes**. To soften the lemons, try rolling the fruit. Roll an orange along with the lemon. Compare the two. Squeeze the lemon and make a drink. ($2\frac{1}{2}$-6)

- 3. Decorate a treat with **almonds** or **whole pecan halves**. (3-6)

- 4. Use **plum tomatoes** to make a salad, salsa, or spaghetti sauce. Offer 2 different shaped tomatoes from which to select-1 oval 1 and round. (3-6)

- 5. Cut ovals from skin colored construction paper and use them to make portraits of family members. (3-6)

- 6. Select potatoes that are oval in shape. Cut the potatoes length-wise into slices, season, bake, and serve for dinner. (2-6)

- 7. Try rolling a football. Compare it to a baseball. "Why do they roll differently?" ($3\frac{1}{2}$-6)

- 8. Sort and count **jellybeans**. (3-6)

- 9. During the spring, decorate Easter eggs and have an egg hunt. Plastic eggs are also available during this season. Look for after Easter sales when they are available for ½ price. You will find many uses for these eggs. Try counting and sorting the plastic eggs. (3-6)
- 10. Crack walnuts for snack. (3-6)

- 11. Bake an **oval pizza.** (2-6)

- 12. Make a large family sized sandwich from a whole **loaf of oval shaped bread** (Un-sliced). (2-6)

- 13. Make **oval shaped hamburgers.** (2-6)

- 14. Use **oval shaped sponges** for cleaning. (2-6)

- 15. Discover that the earth travels in an **elliptical** path around the sun. (4-6)

3-D Shapes
3-dimensional or solid shapes. These

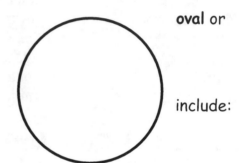

oval or

include:

- Spheres (2-6)
- Cubes (3-6)
- Cylinders (3-6)
- Cones (2-6)
- Pyramids (3-6)
- Rectangular prisms (3-6)

Spheres (2-6)
Vocabulary Development
- Sphere
- Solid
- Circle
- Circular shape

Parent Background Spheres (2-6)
Vocabulary Development
- **Spheres** are solid circles, like a ball. Spheres are round on all sides and will roll when on a flat surface. A flat circle such as a plate will only roll only when it is set on its curved side.
- Use a ball to show your child how he/she can put his/her hands on all of the sides of a **sphere.**

- **A sphere is always a circle; however a circle is not always a sphere.** Only through experiences using both spheres and circles will your child learn to distinguish between the two.

*Going on a Sphere Hunt-Find-

Soccer Balls   Melon Balls   S(
Golf Balls Basketballs   Cotton Balls   A
Baseballs   Green Peas Cantaloupes   M
Beach Balls   Globes   D(
Crystal Balls Meatballs   Pearls   P(
Oranges   Beads
 Grapefruits   Snowballs Grapefruits

Suggested Sphere Activities-(2-6)

- 1. Collect **balls** of different sizes. (2-6)

- 2. Make **melon balls** from cantaloupes or watermelons for a snack. (2-6)

- 3. Boil an egg and examine the **sphere shaped yolk**. Mash the yolk to make deviled eggs.

- 4. Make round sphere shaped **meatballs**. (2-6)

- 5. Roll fresh bread dough into small spheres and bake. (2-6)

- 6. Eat **donut hole balls**. (2-6)

- 7. Go **bowling**. (4-6)

- 8. Inflate a **beach ball**. (2-6)

- 9. Use a string to **measure** all sides of a sphere. "What did you find out?" (3-6)
- 10. Cut an orange into slices and examine the circles. "Can you put the orange back together to make a sphere?" (3-6)

- 11. Use play dough to make spheres, squish each into flat circles. (2-6)

- 12. After a snowfall make a **snowman** or **snowballs**. (2-6)

- 13. Glue craft **pompoms** together to make caterpillars. (3-6)

- 14. Observe the sphere shapes of the **moon** and the **sun**. (2-6)

- 15. Make **ice cream balls**. (Spheres) (2-6)

- 16. Make **blueberry** muffins. (2-6)

- 17. **Roll yarn** into a ball. (sphere) (2-6)

- 18. **Blow bubble** from bubblegum. Use sugar-free gum. (2-6)

- 19. Blow soap bubbles. (2-6)

- 20. String **round pearl beads** to make a bracelet or necklace. (3-6)

- 21. Find **white fluffy dandelions** and blow the seed heads. (2-6)

- 22. Decorate a **Styrofoam ball** with glitter or sequins. Add a ribbon or string and hang. (3-6)

- 23. Make **potato balls**. Mix mashed potatoes, eggs, onion and enough flour to hold mixture together. Roll into balls and drop into hot oil and brown. (2-6)

- 24. Make small **cheese balls** for snack. (2-6)

- 25. Use **cotton balls** to make a lamb or a snowman. Draw an outline of a lamb or a snowman. Color the lamb's face with markers. Glue cotton balls on the body. Cover the snowman's body. Use your imagination to create other things. (3-6)

- 26. Make **popcorn balls**. (2-6)

- 27. Learn the names of the **planets**. Discover the smallest and the largest. ($3\frac{1}{2}$ -6)

- 28. Eat **green peas**. (2-6)

- 29. Enjoy a sphere shaped **lollipop**. (2-6)

- 30. Let your child help to make a salad from **iceberg lettuce** or **coleslaw** from **cabbage**. (2-6)

- 31. Make a fruit salad from cantaloupes, oranges, peaches, (2-6)

- 32. Use a **globe** to locate your state. Locate relatives. (3-6)

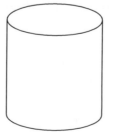

sphere shaped fruit-kiwi, blueberries etc.

state, country, and

- 33. (5-6) Play a game with **marbles**.

- 34. Make a **crystal ball** and use it to tell fortunes. (3-6)

Cylinders-(3½-6)
- -Cylinder
- -Solid
- -Circle
- -Top
- -Bottom

Parent Background- (3½-6)
Vocabulary Development
- All of the following cylinder activities are for children ages 3½-6 only.
- Cylinders are solid shapes that have a circular top and bottom. This shape will roll if put on its curved sides and stand on the two flat circular ends.

- **Go On a Cylinder Hunt**-
- There are lots of cylinders around your house. Open the cabinets and pull out all of those cylinders.

Baby Bottles	Pots	Cheese Sticks
Aerosol Cans	Food Canisters Boxes	Pretzels
Bottles	Paint Rollers	Tootsie Rolls
Straws	Drinking Glasses	Medicine Bottles
Pepper mills	Mugs	Body Parts- Fingers,
Soda Cans	Hot Dogs	Neck, Legs, Arms
A Cut Banana	Sausages	
Cut Carrots	Whole Pepperoni	

Suggested Cylinder Activities-(3½-6)
- 1. Hug a **tree trunk**. Find one that you can fit your arms around. Find a tree trunk that you can put your fingers around.

- 2. Serve large or small **marshmallows** as a snack.

- 3. Cut cylinder shaped fruits or vegetables (carrots, bananas, squash etc.), add to a salad. Let your child select appropriate fruits for this project.

- 4. Freeze juice in **empty juice cans**; add small wooden sticks to make a fun frozen snack.

- 5. Make cookies from a cylinder roll of prepared **cookie dough**.

- 6. Roll dough into cylinders.

- 7. During bath time, use plastic cylinder shaped bottles to explore **volume** and **capacity**. "Can you find one that holds more? Less? Equal?"
- 8. Stack pancakes high into a cylinder.

- 9. Bake a 3-layer cake and stack the circles into a cylinder. "How can we make these cakes into one cylinder?"

- 10. Sort coins and put them into individual **coin holders**, to be taken to the bank.

- 11. String round buttons to make a **cylindrical necklace**. (All of the buttons must be the same size.)

- 12. Make gelatin in empty cans and unmold.

- 13. Make **pencil holders** from empty juice cans. Cut a piece of paper to fit around the can. Draw pictures and glue the paper around the can. Give these holders as gifts.

- 14. Sort the colors in a cylinder roll of candy.

- 15. Use water and a cylinder shaped **paint roller** to paint an outside wall or the sidewalk.

- 16. **Roll paper** into cylinders, to be used as horns.

- 17. Think of things to make from empty paper towel or toilet tissue roll cylinders.

- 18. Slice hot dogs into round 1-inch cylinders.

- 19. Cut a **whole pepperoni** into slices.

- 20. Have a race rolling cylinders. Roll empty salt or coffee cans. Observe what happens. Try rolling the containers up and down a ramp. Race 2 cylinders of the same size. Now try using 2 and different size cans. "What happened?"

-

- 21. Make a cylinder storage box. Use an empty coffee can. Measure a piece of paper to fit around the can. Make a picture and cover the can.

- 22. Make musical shakers. Put rice or beans into an empty plastic cylindrical shaped spice jar or medicine bottles. Have a parade.

- 23. Eat cylinder shaped **push-up pops**.

- 24. Draw with large pieces of sidewalk chalk.

- 25. Draw with colored markers.

- 26. Make **carrot salad**.

- 27. Eat whole pickles.

- 28. Color with **crayons**.

- 29. Cut an empty toilet paper or paper towel roll into 2-inch sections. Use as napkin rings. Color or paint the rings. Make one for each family member.

- 30. Make a flag and attach it to a wooden cylindrical shaped dowel.

- 31. Have fun playing with a **flashlight**.

- 32. Sort cylinder **batteries**.

- 32. Have **cheese sticks** for a 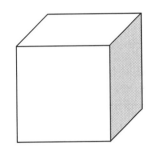 snack.

- 33. Drink using a **straw**.

- 34. Make a drum from a **saltbox** **or a coffee can**.

- 35. Roll your hair with plastic rollers.

- 36. Count the cookies or crackers in a sleeve package. Most cookies come packaged in individual protective sleeves.

Cubes-(3-6)
Vocabulary Development
- Cubes
- Squares
- Solid
- Equal

Parent Background
Vocabulary Development (3-6)
*Cubes are solid squares. All 6 sides are equal.

Mary Taylor Overton

Use the introduction for squares.
Going On a Cube Hunt-Find-
A stack of napkins, ice cubes, a picture cube, dice, houses, blocks, boxes, baskets, tissue boxes

Suggested Cube Activities (3-6)

- 1. Cut a bar of cheese into small **cubes** for a snack.

- 2. Make a book; draw things that are cubed shaped.

- 3. Play games using **dice**.

- 4. Make meatballs into cubes.

- 5. Cut **potatoes into cubes** for dinner. Frozen hash browns are excellent for this project or cut your own.

- 6. Serve frozen peas and carrots. Let your child separate the cubed carrots from the sphere shaped peas.

- 7. Make frozen treats in an ice tray.

- 8. Make a shallow pan of gelatin and cut into cubes.

- 9. Cut **fruits and vegetables into cubes**- apples, melons, cucumbers, or other vegetables.

- 10. Cut cooked meat-ham, beef roast, or cold cuts into cubes.

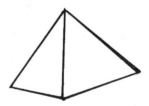

- 11. Slice pound cake and cut into cubes.

- 12. Take a special trip to the store to let your child select a special **cubed shaped** tissue box for his/her room.

- 13. Make stuffing from **bread cubes**.

- 14. Use **sugar cubes** to build or to sweeten a drink.

- 15. Make **croutons** from cubed bread to use in a salad.

Pyramids-(3-6)
Parent Background

Vocabulary Development

- A pyramid is a solid shape. The base has 3 or 4 sides. All of the sides are triangular shaped and come to a point at the top, called the **vertex**.

Suggested Pyramid Activities

- 1. Finding the pyramid shape is difficult. Look for books containing pictures of **Egyptian pyramids**.

- 2. Construct pyramids using straws or or small marshmallows to connect solid

toothpicks and dough shapes. (5-6)

Cones-(3-6)
Vocabulary Development
- Circle
- Circular
- Point

Parent Background
- **Vocabulary Development**
- Cones have a **circular base**. The sides form a point at the top called the **vertex**.
- **Cones**

Finding Cones
➤ Ice Cream Cones
➤ A Flag Pole
➤ Top of a Castle
➤ Birthday Hat
➤ Dunce Cap
➤ Top of a Rocket

Suggested Cone Activities-(3-6)
Roll a piece of paper into a cone.
- 1. Use the cone to serve a treat-chips and popcorn etc. (3-6)

- 2. Pour gelatin into sugar **ice cream cones** and gel. (3-6)

- 4. Construct a castle from empty boxes. Use ice

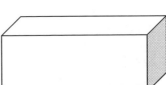

cream cones for the tops of the towers. (3-6)

- 5. Scoop ice cream into sugar cones.

- 6. Make cone shaped party hats. (3-6)

- 7. Turn rolled cones upside-down for a **teepee**. Decorate. Make several for a Native American village. (3-6)

Rectangular Prisms
Parent Background(3½-6)
Rectangular prisms are solid rectangles. See the activities in the **rectangle section**; most can be applied to **rectangular prisms**.

Suggested Rectangular Prisms (3-6)
- 1. Decorate a **brick** to use as a doorstop. Use acrylic paint to paint a design on the brick. To keep from scaring the floor, glue a piece of felt on the bottom. (3-6)

- 2. Paint an empty **shoebox.** (3-6)

- 3. While shopping, let your child select a special rectangular **tissue box** for his/her room. (3-6)

- 4. Wash your hands with a rectangular **bar of soap**. (3-6)

- 5. Store your pencils in a **pencil box**. (3-6)

- 6. Look in the cabinets to find rectangular prism boxes. (3-6)

Lawn and Garden Geometry-(2-6)
Parent Background
The lawn and garden are perfect places for putting geometric observations and skills into practice.
Vocabulary Development-
- -Area- The space within a boundary.
- -Incline- A slight rise in the land
- -Measure
- -Shape
- -Length
- -Width
- -Perimeter-The measurement around the area or shape

Suggested Lawn and Garden Activities (2-6)

- 1. In the yard or garden, measure a space for planting, raking leaves, shoveling snow, cleaning, playing etc. Clearly mark off the **area.** Walk the **perimeter** (the outside) with your child. Discuss what work will be done in this **area.**

- 2. "This is my **area** and that is your area."

- 3. "We will plant flowers in this **area**, from this corner to this corner."

- 4. "Can you show me the area where we will plant flowers?"

- 5. "This is the area of the yard where you may play. Do not go beyond this boundary." Trace around the perimeters of the area.

- 6. "Let's make your **area into a rectangle.**"- A square, circle, or a triangle.

- **7-Use comparison vocabulary-**
- ✓ -Long-longer- longest
- ✓ -Short- shorter-shortest
- ✓ -High- higher- highest
- ✓ -Low- lower- lowest
- ✓ -Tall- taller- tallest
- ✓ -Big- bigger- biggest
- ✓ -Large- larger- largest
- ✓ -Small- smaller- smallest
- ✓ -Narrow-narrower-narrowest
- ✓ -Wide- wider- widest
- ✓ -More- less

- 8. "Is the grass on this side, **(higher or lower)** or **(longer or shorter)** than the grass over here?"

- 9. "Is this area **larger or smaller?**"
- 10. "Can you find the **tallest** flower in the garden?

- 11. " Find something that is **wider?**"
- 12. "Can you find something **taller- shorter- longer?**"

- 13. "Find the **tallest** tree?" "Find one that is **shorter?**" "Find the **shortest** tree?" "Find the **fattest** tree?"

- 14. "Can you find a **triangle, circle,** and a **cylinder** shape?"

Mary Taylor Overton

The bathroom is a wonderful place to explore geometry. (2-6)

- 1. Identify **squares** (floor tiles, wall tiles, wash cloths etc.).

- 2. Provide chalk to make pictures on the **square** tiles.

- 3. Use chalk to connect **square** tiles to make **rectangles**, or **triangles** etc.

- 4. Identify **rectangles**–tub, towels, tissues boxes, soap, 2 squares of toilet tissue, walls, shower stalls, bath mats, medicine cabinets, windows, doors, soap dishes, and toilet tank tops.

- 5. **Circles**-can be found everywhere; the drain in the tub, sink, door knobs, tops of bottles and jars, lights, shower curtain rod, towels, and toilet tissue holders.

- 6.The bathroom is also a good place to discover and use descriptive words.
  Curved   Straight   Angles   Lines   Vertical   Horizontal

Geometrical shapes can be found in all sports

See if you can find any of these shapes?

- Balls-spheres
- Basketball hoops-circle
- Basketball court-rectangle
- Football field-rectangle
- Football-oval
- Goal post-rectangles
- Baseball field-square
- Baseball bat-cylinder
- Ice skating rink-oval
- Tennis court-rectangle
- Tennis racket-oval
- Soccer goal-rectangle
- Boxing ring-square
- Domed stadium-semi-circle
- Swimming pool-rectangle
- Golf- ball cup-cylinder
- Golf bag-cylinder
- Rule book rectangles
- Race track-oval
- Bowling ball-sphere

Numbers and Counting
Parent Background-(2-6)

- Our number system is a **language** in itself, so think in terms of helping your child to master all of the components of this language.
- The understanding of numbers and counting skills take time to develop and evolve.
- Take every opportunity to use numbers and counting.
- Numbers and counting go together and should be made an important part

of your child's everyday experiences.
The object is to help the child to make connections that-

Number Objectives (2-6)

-To discover numbers in the child's world- Numbers give you information
-To count- Counting is sequential and follows an order
-To identify number symbols
-To make number sets
-To identify sets
-To match sets and numbers

- The child must understand the relationships between-

-1. The number name (**spoken two**)
-2. The number symbol (**2**)
-3. The number of **objects**-that two represents 2 single object in a given set- A set is a group of objects- (X X) = 2. (1-to-1 correspondence- each object represents 1 member of the group)

- To master numbers the young child must be given **step-by-step** number activities that are:
  1. **Sequential-follow an order**
  2. **Age-appropriate**
  3. Presented in ways to **maximize understanding** of the total picture

- The first steps for (2-6) is to:
✓ Learn to recognize number symbols by sight
✓ Learn to Count
✓ Learn quantities

• Rote counting

- Rote counting is saying numbers in sequence (1-2-3-4) and is an important component of math.
- Your child may be able to recognize number symbols and rote count from 1 to 25 or beyond, and still he/she may not have an understanding of numbers.
- Do not depend on rote counting as the only measure of your child's number understanding.
- When rote counting, count slowly. Pay attention to the numbers 15 and 16. Some children have difficulty hearing and distinguishing between 15 from 16. These two numbers sound alike, and many children often skip one.

• Understanding quantity

- Without the child having some understanding of **quantities**, rote counting and symbol recognition do little more than to put numbers in sequence.
- **When counting objects it is very important that your child touch each object as he/she says the number.** This gives the idea that each objects stands alone as one unit in itself-(1-to-1 correspondence).

Mary Taylor Overton

Make sets
- As your child groups five objects he/she is learning to recognize and understand relationships between amounts, number symbols, values, comparisons, and number names.

Compare sets
- By comparing the number of objects in one set to the number of objects in another, the child is given an opportunity to visually discriminate differences between sets of different quantities.
- This helps him/her to make connections between numbers and quantities

- -To understand 1 to 1 correspondence
- -To compare number values

Numbers play a vital role in our lives, so help your child to discover numbers in his/her daily life. **Go on a number hunt.**

Some Suggestions Where Numbers Are Found-

Addresses	Money	Route Numbers
Channels	Calendars	Radios
Telephones	Ages	Credit Cards
License Plates	Clothes-Shoe Size	Checks
Medications	Weight	Police Cars
Newspaper Ads	Speed Limits	Taxis
Thermometers	Buses	Airplanes
Grocery Stores	Sport Events Score	Computer Keyboards
Weather	Book Pages	
Gas Stations	Calculators	
Recipes	Gas Station Pumps	

- 1. **Let your child see you using numbers**- writing a check, banking, checking grocery store prices, counting money, measuring, checking the date, using your watch, etc. (2-6)

- 2. Magnetic numbers on the refrigerator are excellent for teaching number recognition. Make your own. Write numbers on heavy card stock paper and glue magnets on the back. Strips of magnetic tape are available at any office supply store. As your child progresses add more numbers.

- 3. Cut large numbers from **colored contact paper** and display them on the wall in your child's room.

- 4. **Make flash cards.** (2-6)Use these cards in conjunction with rhymes, learning to read numbers, number stories, and making sets etc. On large index cards make number cards 0-20. Let your child decorate the back.

Use the number cards with the number being learned. New numbers should be added as your child progresses. Don't forget about the **number 0.** This number will have more meaning when he/she gets to **place value**, and **tens and ones.** He/she will encounter these skills in late kindergarten or 1st grade. An example: the number 29 is equal to **2 tens** and **9 ones** (2 sets, each having 10 members and 1 set with 9 members). Combined, these sets are equal to **29.**

Suggestions for Introducing Numbers (2-6)

- 1. Introduce 1 and 2 by displaying the numbers. When you feel that your child has mastered these two numbers, add one more. **Remember take your time**, you may be ready to move on, but your child may not.

- 2. Have a "**2's Dinner.**" There must be 2 of every item served on the plate. (Cut hot dogs into 2 pieces, 2 meatballs, a potato cut into 2 pieces, 2 pieces of broccoli, 2 slices of tomatoes, 2 carrot sticks or celery sticks) (2-6)
  - ➢ Serve yourself 2 spoonfuls of rice.
  - ➢ Use 2 forks-1 salad, 1 regular.
  - ➢ Use 2 straws, at the same time.
  - ➢ Make a fruit salad-2 chunks of bananas, melon balls, and pineapple chunks.

- 3. Count by touching each object as you count- 1-2 blocks. (2-6)

- 4. Identify body parts that come in 2's-arms, hands, feet, ears, knees, legs, cheeks, and lips etc. (2-6)

- 5. "You may take 2 stuffed animals to bed." "Can you count them?" (2-6)

- 6. Count **backwards** 3-2-1-0 (2-6)

- 7. Brainstorm things that come in 2's. (Earrings, gloves, mittens, socks, shoes, curtains, gloves, glasses etc.) (3-6)

- 8. This is a perfect time to introduce the term **pair**. Talk about the meaning of a pair of items. Find things that come in a pair. (2-6)

- 9. Introduce **sorting** and **matching.** Sorting is an **organizational skill.** From your collection of buttons select 2 matching buttons; you will need at least 6 matching sets. Place the sets together in a separate box. Your child is to sort through the buttons to find each of the 2-matching. Also try using coins, candy, Legos, pasta, keys etc. (2-6)

- 10. During bath time, to the water add sets of 2 small matching objects. The object of the game is to find the 2 matches. Examples: pennies, bars of soap, medicine bottle caps, any floating toys, spoons, straws, or small pieces for an old game. (2-6)

- 11. Play a game of "**Simon Says.**" Give your child simple directions-
  - Clap 2 times.
  - Stomp 2 times

Jump, stomp your foot, snap your fingers, point your finger, tap your toes, pull your ear, blink your eyes (count out loud). (2-6)
Extend "**Simon says.**" Before performing the task, your child must first identify the number on the card. "Simon says, jump 4 times." (2-6)

- 12. Use a clock to count **clockwise** 1-12. As you count, use your finger to point to the numbers. (2-6)

- 13. Count **counterclockwise** 12-11-10-9-8-7-6-5-4-3-2-1-0. (2-6)

- 14. Count blocks as you stack them. (2-6)

- 15. While on a walk, count the sidewalk squares. "This time I am going to say number 1. As you step on the next square, you must say the next number?" (2-6)

- 16. Play **hopscotch**! "Look! You just hopped on the number 2! What number are you standing on now?" (2-6)

- 17. "Count the number of people at the dinner table." (2-6)

- 18. "How many pieces of silverware are on the table?" (2-6)

- 19. Pick-up a handful of pasta. Count the pasta and cook for dinner. (2-6)

- 20. "How many green beans are on your plate?"

- 21. "How many letters are in each family member's name?" (2-6)

- 22. "How many buttons are on your shirt?" (2-6)

- 23. "How many jellybeans are in your hand?" (2-6)
- 24. "Let's count the toys as we put them away." (2-6)

- 25. "How many markers are in the box?" "Let's count as we put them away." (2-6)

Use the following activities to learn how to make numbers. ($3\frac{1}{2}$-6)
- Say the chants as you make the numbers in the air. (2-6) 19. **Play number**

games. Whoever can find 3 socks first, wins! Whoever can clean up 8 blocks first, wins! (2-6)

- 27. Counting often happens in the kitchen. "Add 2 eggs." "Would you like 3 or 4 chicken nuggets for lunch?" "Please have 2 more bites." "How many cookies would you like?" (3-6)

- 28. When emptying the dishwasher count the forks as you put them away. Count spoons, cups, and plates.... (2-6)

- 29. Play games that use number spinners! (2-6)

- 30. Use a timer to time a task. (3-6)

- 31. Set a timer for 3 minutes and have everyone in the house try to find something with the number 2 on it. Try using other numbers. (3-6)

- 32. Count the number of steps it takes to walk down the hall. (2-6)

- 33. When reading a book, say the **page numbers**. "Now we are turning to page number 5...." Use your finger to point to the number. (2-6)

- 34. The **telephone** is an excellent way to teach number recognition. Remember to give one number at a time. It is important for your child to learn his/her **home telephone number**, **your work number**, and of course **911**.

- 35. Make your own **telephone** and **address book**. (4-6)

- 36. Teach your child his/her **address**. Learn to recognize the numbers. (4-6)

- 37. Read **number books**. Most of these books have numbers as well as corresponding sets of objects. Say the number and count the objects. (2-6)

- 38. **Make your own number books**. Put a different number on each page. Draw or paste a set of objects to match the number. Make up a story to go with each page. (3-6)

- 39. Count beads as you string them to make a necklace. (3-6)

- 40. For holiday gifts, purchase board games that require counting. (2-6)

- 41. As you are preparing dinner, count the ingredients needed for a recipe. (2-6)

Mary Taylor Overton

Number Rhymes
- 1. At a very early age, play number and counting rhymes. Look for every teachable moment to re-enforce number and counting skills. Young children enjoy rhymes that are lively and catchy. Try creating your own. ($1\frac{1}{2}$-6)
- 2. Fingers and toes are always available for counting, so take full advantage of them. ($1\frac{1}{2}$-6)

Suggested Number Rhymes-
Rhymes can be used as a motivation for introducing numbers.
Number 1
- **Hickory Dickory Dock**
Hickory dickory dock
The mouse ran up the clock.
The clock struck 1.
And down he ran.
Hickory dickory dock.

Number 2

- **Hot Cross Buns**
Hot cross buns!
Hot cross buns!
1 a penny
2 a penny
Hot cross buns!
(Use 2 pennies to count 1-2)

- **2 Blackbirds**
There were 2 blackbirds
Sitting on a hill.
One named Jack
One named Jill.
0-1-2

Fly away Jack!
Fly away Jill!
Come back Jack!
Come back Jill!
0-1-2

- **Song- Tune- Mary Had A Little Lamb**
2 little hands go clap, clap, clap.
Clap, clap, clap.
Clap, clap, clap.

2 little hands go clap, clap, clap.
Come and hear them clap, clap, clap.

(Other verses)
- 2 little feet go stomp, stomp, stomp.
- 2 little eyes go blink, blink, blink.
- 2 little knees go bend, bend, bend.
- 2 little lips go kiss, kiss, kiss.
- 2 little arms go shake, shake, shake
        (Make up your own verses)

Number 3

- **Three Little Kittens**
Three little kittens have lost their mittens.
And they began to cry.
Oh mother dear we sadly fear,
That we have lost our mittens.
Oh dear! Do not fear,
My little kittens,
And please don't cry.
For I have made a pie.

- **Baa Baa Black Sheep**
Baa baa black sheep.
Have you any wool?
Yes sir. Yes sir.
Three bags full.
One for my master.
One for the dame.
One for the little boy who lives down the lane.
(Point to your fingers and count) 1-2-3.

Read books about 3's
        3 Little Pigs
        3 Bears
        3 Billy Goat Gruff
The Three Bears – count bowls, chairs, beds, spoons etc.
The Three Billy Goats Gruff-green, greener, greenest
The Three Little Pigs- strong, stronger, strongest

Number 4
- **Ready Set Go**
1 for the money
2 for the show

3 to get ready
And 4 to go.
(Use this for starting a race anything- count backwards 4-3-2-1-0)
1054

Number 5
This Little Piggy Went to Market
This little piggy went to market.
This little piggy stayed home.
This little piggy ate roast beef.
This little piggy had none.
This little piggy cried,"Wee, wee, wee," all the way home.
  (0-1-2-3-4-5 touch each finger again)

Number 6-Have fun finding rhymes that use the number 6

Number 7
Cut out 7 potatoes from brown paper or use real potatoes.
• **One Potato**
1 potato, 2 potatoes, 3 potatoes, 4.
5 potatoes, 6 potatoes, 7 potatoes more.
(0-1-2-3-4-5-6-7)

Number 8-Have fun creating your own rhymes

Number 9

• **Engine Engine Number Nine**
Engine engine number nine
Running on Chicago line.
If the train runs off the track will you get your money back?
0-1-2-3-4-5-6-7-8-9.   9-8-7-6-5-4-3-2-1-0

Number 10
• **Where Is Thumpkin**
Where is thumpkin?
Where is thumpkin?
Here I am Here I am.
How are you this morning?
Very well I thank-you.
Run away run away.

2-Where is pointer?
3-Where is tallman?

4-Where is ringman?
5-Where is pinky?
6-Where are all the men?
  Where are all the men?
  Here we are. Here we are.
  How are you this morning?
  How are you this morning?
  Very well I thank-you.
  Very well I thank-you.
  Run away run away.
  0-1-2-3-4-5-6-7-8-9-10

- **This Old Man**

This old man he played zero
 He played nick-nack on my hero.
 With a nick-nack patty-wack give the dog a bone.
This old man came rolling home.
This old man he played one.
He played nick-nack on my thumb.
With a nick-nack patty-wack give the dog a bone.
This old man came rolling home.

(Other verses)
This old man he played 2.... on my shoe.
This old man he played 3..... on my knee.
This old man he played 4..... on my door.
This old man he played 5.... on my side.
This old man he played 6.... on my sticks.
This old man he played 7...   up to heaven.
This old man he played 8.... on my gate.
This old man he played 9.... on my spine.
 This old man he played 10... once again.
0-1-2-3-4-5-6-7-8-9-10

- **Song- Tune 10 Little Indians**

1 little, 2 little, 3 little fingers.
4 little, 5 little, 6 little fingers.
7 little, 8 little, 9 little fingers.
10 fingers wiggle in the air.
0-1-2-3-4-5-6-7-8-9-10

- Counting backwards (4- 5)

10 little, 9 little, 8 little fingers.
7 little, 6 little, 5 little fingers.

Mary Taylor Overton

4 little, 3 little, 2 little fingers.
1 little finger falling down.
10-9-8-7-6-5-4-3-2-1-0

- Play rocket blast off by counting backward.
  10-9-8-7-6-5-4-3-2-1-0 blast off. (2-6)

Counting On(2-6)

Counting on is a skill that takes time to master. You say 0-1-2; your child must begin counting from where you left off, 3-4-5. (3½-6)

Skip Counting

Skip Counting is a way of counting. (5-6) It is often used by children in games and chants. It uses numbers in rhythmic patterns, such as counting by 2's, 3's, 4's, 5's, 6's, 7's, 8's, 9's. and 10's. Skip counting by 2's- (0-2-4-6-8-10-12-14-16) Skip counting is a fun way of rote counting. Start by counting by 10s. (4-5)

Ordinal Numbers-

This is another way of **sequencing numbers or events**. (2-6)

- First
- Second
- Third
- Fourth
- Fifth
- Sixth
- Seventh
- Eighth
- Ninth
- Tenth
- Last

- When discovering **ordinal** numbers, begin slowly. Your child must understand the meaning of **first**, **last**, **before**, **next**, and **after**.

Objectives-
- To understand when the use of ordinal numbers is appropriate
- To use ordinal numbers as a sequencing tool

- ## Suggested Ordinal Activities (2-6)
- 1. Give simple **2 step directions**. (2-3½)
"**First**, you must take your bath, second, brush your teeth."
    (4-6) can handle directions that have **3-4 steps**.

- 2. "Before dinner you must **first** wash your hands."

- 3. Follow simple directions to play a game, to draw a picture, or to follow a recipe etc.
  Example-
✓ **First**- put all of the pictures down.
✓ **Second**- match the two sets of pictures.
✓ **Third**- Put each set of matching picture over here.

- 4. "Put the blocks in order." "The red block is 1st, blue block 2nd, and green block 3rd."

- 5. Place stuffed or plastic animals in a line.
  Play a game. "Give me the third animal." "Show me the 1st animal."
"Touch the 2nd animal." "Put the 4th animal under the table."

- 6. "Stop on the 2nd step."

- 7. "Jump on the 3rd sidewalk square."

- 8. When you eat the **last** French fry say, "All gone."

- 9. 1st wash your hands, 2nd get your coat, 3rd you may go outside.

- 10. List the **sequential order** to be followed to make something or perform a task. (4-5)

- # Sequential Order-
❖ Make a bed.
❖ Make toast
❖ Brush your teeth
❖ Get ready to take a bath
❖ Get ready for school
❖ Make a sandwich

- 11. Place 5 blocks of different lengths in order of height. Put the tallest one first. ($3\frac{1}{2}$-6)

- 12. Cut straws into varying lengths. Your child is to order the straws according to of lengths- (short to longest, or long down to the shortest. (2-6)
- 13. Have fun by changing, "**This Little Piggie Went to Market**" into ordinal numbers. (4-6)
The **first** little piggy went to market.

Mary Taylor Overton

The **second** little piggy stayed home.
The **third** little piggy had roast beef.
The **forth** little piggy had none
And the **fifth** little piggy cried,
Wee, wee, wee all the way home.

- 14. Read the stories **Three Billy Goats Gruff, The 3 Bears, The 3 Little Pigs** and substitute ordinal numbers. (3½-6)

By the age of 3 ½ some children's fine motor skills have developed enough to introduce writing numbers on paper. **Never make him/her just sit and practice making numbers.** Use chalk, crayon, paint, finger paint or shaving cream. **This is a fun activity, so approach it as just that!**

- The number 1- Slide right down.

- The number 2- Around and down and then back over.

- The number 3-Around and stop around and stop.

- The number 4- Down and over and then slide right down.

- The number 5- Down, add a big fat belly and then a cap.

- The number 6- Down and around and bring up the foot.

- The number 7-Over and slide down the mountain.

- The number 8-Make an S and go to the top.

- The number 9- A ball and a bat

- The number 10- slide right down and a zero.

Addition and Subtraction (3-6)
Parent Background
- Number facts are the corner stones of math. (2+2=4) (5+3=8)
- A child must have instant recall of these facts.
- Number facts are built of sequential patterns.
- The goal is to help the child to **develop strategies** for understanding these patterns.
- You want your child to understand that there are many different ways to solve a problem.
- To do this he/she must develop **problem solving strategy skills**.
- Problem solving skills enable the child to look at a problem in a variety of different ways. Strategies give the child alternative paths to finding a solution to the problem.
- Always present problems as fun, hands-on activities, using actual objects, and real-life situations. Your child will be solving problems and learning math number facts without realizing that he/she is doing so.

Always:
- ✓ Look for opportunities to present problems.
- ✓ Give your child time to think about his/her solution.
- ✓ Provide manipulatives, so that he/she can see and touch.
- ✓ Talk about the solution.
- ✓ Look for other ways to solve the same problem.
- ✓ Never, drill.
- ✓ Make problems that are age-appropriate and in keeping with your child's skill level.
- ✓ Never move on too quickly.

The following are suggested examples to show how to construct math story problems. Your child will need to engage in many activities just like the examples below. It is your job to look for opportunities to create more problems that require him/her to combine sets. **Always keep in mind the skill level of your child.**

Suggested Addition and Subtraction Activities
(2-6)

- 1. "You have 2 blocks. If I give you 2 more, how many do you have?"

- 2. "If I give you 3 pennies, and then 2 more, how many pennies do you have?"

- 3. "Let's set the table, here are 2 plates, get 3 more. How many plates are there?"

- 4. "I have 4 buttons; give me 3 more buttons. How many buttons do I have?

- 5. "You have 6 buttons; give me 2 buttons. How many buttons are left?"

- 6. "There are 5 flowers in the garden. You may pick 2 flowers. How many flowers are left?"

- 7. "We are baking a cake; we need 3 cups of flour. We have 1 cup. How many more cups do we need? "

- 8. "Count the number of french fries on your plate. If you eat 2, how many French fries do you have left?"

- 9. "You have 5 coins. If I give you 3 more, how many do you have?" "Here are 2 more. Now, how many coins do you have?"

Division and Multiplication (4½-6)
Multiplication and division can also be introduced without your child realizing it.

Multiplication
- 1. "There are 4 people to eat lunch; each person needs two pieces of bread. How many pieces of bread do we need?" Give your child time to think how to solve this problem.
 Possible answers:
 ✓ Count 4 pieces of bread and then place another piece on top of each.
 ✓ Put out 4 groups of 2 pieces of bread.

- 2."There are 2 children. You must give each child 4 cookies. How many cookies do we need?"

- 3."There are 3 glasses. You must put 2 pieces of ice in each glass. How many pieces of ice do you need?"
DIVISION
- 1. "You have 6 hats. There are 3 children, how many hats can each child have?"
Possible Answer:
 ✓ 1. Give each of the 3 children, 1 hat at a time and repeat until all 6 hats have been given out.

- 2. "I have 12 pieces of bread. How many sandwiches can I make?"

- 3. "Here are 10 cookies. There are 2 children, how many cookies will each child get?"

Measurement

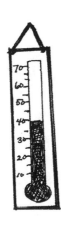

Measurement(2-6)

Parent Background

- Standard measuring tools are instruments used by man to keep the world and his/her life in a unified order.
- These tools insure that we are all working on the same page.
- Each tool is **designed to perform a specific task**.
- Each has its **own number unit value base**.

- **Time measurement-**
- **(Clocks)**-second, minute, hour
- **(Calendars)**-day, week, month, year, decade, century

- **Volume measurement**-cup, pint, quart, gallon, teaspoon, tablespoon,
  liter, bushel
- **Liner measurement-(Ruler)**-inch, foot, yard, meter, mile

- **Temperature measurement-(Thermometer)**-degree, Fahrenheit, Celsius

- **Estimation measurement**-a guess based on prior knowledge

- **Mass measurement-(Scales)**-weight-ounce, pound, ton

- **Money measurement**-penny, nickel, dime, quarter, ½ dollar, dollar

- **Speed measurement-(Odometer)** miles per hour, light year

- To use any one of these **standard-measuring tools** there must be an understanding of:
➢ What the tool is
➢ What kind of information can be gathered
➢ How to use the tool to gather information
➢ The value of each unit
➢ How to apply the information
➢ How to make use of these instruments in our daily lives

- All **standard measuring tools,** clocks, rulers, calendars, odometers, scales, and thermometers are **number based** and are used to gather information.
- After the information is gathered, one must be able to understand how to

apply this information both **concretely** and **abstractly** (mentally).

- Because young children lack the abstract thinking skills needed to process and apply the measurement information they have difficulty with standard measuring tools.
- To think abstractly, the child must have some basic understanding of the tool and its use. He/She must also have a **frame of reference base** (from prior experiences), to apply, evaluate, and understand the information.
- An example is the **thermometer**-the tool and **temperature**-the number of **degrees** it is meant to **measure** and **show**. When an adult hears that the temperature will be 85 degrees, from prior knowledge about how 85 degrees feels, he/she immediately understands what this information will mean to his/her life. Adults are also able to conclude when the temperature has/has not reached approximately 85 degrees or more.
- 2's-5's also have difficult understanding how to interpret the individual units upon which each standard measuring tool is based. For example-learning that 60 seconds is equal to 1 minute, is **knowledge in isolation**, until it is applied to the total concept of time. 60 seconds takes on new meaning when there is an understanding that:
  - ➢ 60 seconds = 1 minutes
  - ➢ 60 minutes = 1 hours
  - ➢ 24 hours = 1 day
  - ➢ 7 Days = 1 week
  - ➢ 4 weeks = 1 month
  - ➢ 12 months = 1 year
  - ➢ 10 years = 1 decade
  - ➢ 100 decades = 1 century
- Adults are able to think of these units of time collectively as well as separately; abstractly and concretely as they apply this information to the situation.
- However, even with this information, young children have difficulty understanding and applying time to their daily lives.
- Only through time, age-appropriate experiences, and connections will the child's **concrete** and **abstract** thinking merge into an understanding of how to apply time measurement information.
- Abstract thinking is **developmental** and takes time to evolve.
- By the age of 6, your child's cognitive skills (thinking) and his/her past experiences should allow him/her to **begin** to use standard-measuring tools, with **limited skill**.
- This is not to say that you should wait until your child reaches the age of 6 to begin learning about measurement.
- **By the age of 2**, start laying the foundation for understanding basic measurement concepts.
- (2-4) You want him/her to understand that each of these tools: Clocks, rulers, calendars, thermometers, scales etc.
  - ✓ Plays an important role in his/her life

- ✓ Performs a certain task
- ✓ Is based on specific units
- ✓ Is used to show a quantity
- ✓ Follows a sequence
- **Do not be tempted to jump ahead.** Any child can be taught to read or say- 5 inches, $3.00, 4:00 etc. However, without the child having some understanding of the measurement process-inches, dollars, and o'clock will have little meaning.
- (2-4)Focus on providing lots of non-standard measuring **comparison experiences, explorations,** and **developing an age appropriate comparison vocabulary.**
- Help your child to understand the meaning of such comparison concepts as, Which is **taller?** Which is **larger?** Who is **older?** Which will hold **more? Today** is…. What is **tomorrow?** Use anything with in the child's world. "Can you use these blocks to make a **tall** tower?" "How can you make it **taller?**" "Can you **add more** to make it taller. "Let's **count** to find **how many?**" **Do not be tempted to jump ahead.**
- By the age of 4, your child's counting and number awareness skills should be strong enough to gradually begin attaching numbers to the comparisons. "There are 5 blocks on this tower and 3 on that one, which one has more? Let's match each block 1-to-1 to see which one has more."
- Around the age of (5-6), **gradually** begin to use **standard measurements-**feet, hours, inches etc. **A word of caution,** even by this age **many** children are not ready to handle these concepts with accuracy.
- **Continue comparisons.**

Objectives-
1. To learn related vocabulary
2. To identify some standard units of measurement
3. To use comparisons as a measuring tool
4. To understand how measurement can help you to gather information
5. To experience time
6. To explore non-standard measurement as measuring tools ($3\frac{1}{2}$-6)

Time-(2-6)
Parent Background-
- **Time** can be thought of as **clock time** and **calendar time**. We use them to put our lives into an orderly sequence.
- Each is used to show the exact **time** now, the **elapse of time,** and the **time that is to come.**(time left in a given unit of time)
- When a 3 or 4-year-old child is put in, **"Time Out,"** for fifteen minutes, he/her equates it as being an hour or two. This is because in relationship to the actual length of 15 minutes **he/she has no frame of reference.**

He/she has not yet formulated a base on which to compare the actual time that has elapsed.

- Adults have had many experiences with fifteen-minute segments of time. They mentally (abstractly) know approximately how long it takes for the fifteen minutes to elapse.
- Adults can think of an hour as, four- 15-minute segments, sixty- 1 minute, or two-30 minute intervals. They are also aware of how long it will take to perform certain tasks. Throughout the day adults are always aware of the approximate time of the day. Through life experiences adults have internalized an array of **different** time comparison bases.
- They are able to apply these abstract bases in conjunction with the concrete clock and calendar measuring instruments.
- To help your child to **build his/her own time comparison bases**, he/she must be given **pre-clock** and **pre-calendar** comparison experiences.
- His/her understanding of how to apply time happens as he/her experiences time in relationship to his/her daily activities.
- 2's–6's should not be expected to tell time with accuracy.
- At an early age the child needs to understand that:
✓ **Time plays a major role in his/her life.**
✓ **Clocks give you information.**
✓ **Calendars and clocks are used to keep track of time.**
✓ **Time is sequential.**
✓ **Time has a definite length.**
- By the age of 6, children should have made enough connections with the abstract and the concreteness of time to formally **begin** learning to tell time. Even at this age his/her time accuracy will be **very limited**.
- Time is taught from the largest unit of measurement down to the smallest. Example-Time on the hour is the easiest unit for young children to understand. The hour is followed by the 1/2 hour, quarter of hour, minutes and down to seconds

Objectives-Clock Time (2-6)
1. To use related vocabulary
2. To understand that clocks are used to measure a span of time
3. To understand that time is sequential
4. To understand time is composed of units of value
5. To understand that we do different things at certain times of the day

Vocabulary
(2-6)
✓ Clock
✓ Before

- ✓ After
- ✓ Morning
- ✓ Breakfast

(3½-6)
- ✓
Afternoon
- ✓ Lunch
- ✓ Evening
- ✓ Dinner
Night

(4½-6)
- ✓ Hours
- ✓ Minutes

Take a look around your house to find and count the number of **clocks**. They can be found on the VCR, computer, cable box, microwave oven, clock radio, etc. Nine out of ten clocks in the average household are **digital**. It is difficult to teach young children how to tell time using a digital clock or watch. Without an understanding of time, digital time instruments offer young children **little more than an exercise in reading numbers**. Go out and purchase an old fashion clock with a face and hands.

Suggested Time Activities-(2-6)

- 1. Give your child a timeline of his/her day. This gives **a frame of reference**. (2½-6) If you plan to pick up him/her up early from school, always give the order of the day. "I will pick you up after rest time, before snack." Do not just say, "I will pick up you up early." Early to the young child means just that, so he/she spends the day thinking that early could be any minute.

- 2. Let your child see you using a **watch** and **clocks**. (2-6)

- 3. Give your child a **watch** to wear. He/she may not be able to tell time, but it will give a sense of the **relationship** between the **instrument** and **time**. (3-6)

*Use age appropriate vocabulary.
(2½-3) **Morning, noon,** and **night**.
- 1. We wake up in the **morning**.
- 2. We eat **lunch** in the **afternoon**.
- 3. We eat **dinner** and go to bed at **night**.

(3½-4) **morning, noon, afternoon, evening, nig**ht
1.   We wake up in the **morning.**
2.   We eat **lunch** at **noon.**
3.   We play in the park in the **afternoon.**
4.   We eat dinner in the **evening.**
5.   We go to bed at **night.**

(4-6) **Morning, noon, afternoon, evening, night**
1.   We wake up at **7:00,** in the morning.
2.   We eat lunch at **12:00,** noon.
3.   We play in the park at **4:00,** in the afternoon.
4.   We eat dinner at **6:00,** in the evening.
5.   We go to bed at **8:00,** at night.

- 4. (3½-6) Use a clock to indicate the time of events, on the hour 6:00; **never** try 6:25 or 7:45. This is a skill that will come when he/she has a better grasp on the concept of time and its relationship to the clock.

- 5. Put up a clock in your child's room. Make a clock, using a paper plate. Add movable hands. Use the clock to show bedtime or other events. (3½-6)

- 6. Get a two-minute egg **timer** and a wind-up timer and use for timing activities. (2-6)

- 7. **Measure Time.** "How many things can you pick up before the egg timer runs out? 1 minute?" (3½-6)

- 8. "Let's see how long it will take to eat lunch." (2-6)

- 9. Set the timer for taking a bath, getting dressed, or brushing teeth. (2-6)

- 10. **Estimate time** (5-6) Snap your fingers and tell your child that a second is a measure of time, quick like a snap of the fingers. Watch the second hand travel once around the clock and explain that's the length of a minute. Explain that an hour is approximately the time that it takes to take a bath and read a goodnight story.

- 11. Count in seconds, by saying, "1 one thousand, 2 one thousand, 3 one thousand." It takes approximately 1 second to say each number. Have fun timing activities using seconds. (5-6).

- **Things that you can do in an estimated second**

✓ Clap
✓ Stomp
✓ Cluck your tongue
✓ Wink
✓ Jump

✓ Sneeze
✓ Snap
✓ Kiss....
✓ Twist
✓ Turn around

- **Things that you can do in an estimated minute**

In an estimated you can...

✓ Sing a song
✓ Make a sandwich
✓ Pour juice
✓ Wash your hands

✓ Peel a banana
✓ Hop around the room
✓ See a TV commercial
✓ Set the table

- **Things that you can do in an estimated hour**

✓ Watch two- 30 minute kid's videos
✓ Go the grocery store
✓ Take a long walk

✓ Bake a cake
✓ Eat dinner with the family

Calendar-Time (2-6)

Objectives-

1. To learn appropriate vocabulary
2. To understand that a calendar is used to keep track of the days of the week and the months of the year
3. To understand that days, months and years are sequential
4. To understand that certain events happen on different days.

VOCABULARY-

- Calendar
- Day
- Days of the week
- Week
- Month
- Months of the year

- Year
- Older
- Younger
- Yesterday
- Today

Tomorrow

Suggested Calendar Activities (2-6)

- **Parent Background**
- Place a **large calendar** where it can be seen.
✓ Each morning call attention to the calendar.
✓ Name **yesterday, today, tomorrow.**

✓ If your child is in pre-K through first grades, he/she will also do this same exercise at school. You are giving him/her some **prior knowledge** or a **frame of reference** that he/she may apply at school.

- 1. Mark off the day on the calendar. (3½-6) Discuss the month, day of week, year, and any special event.

- 2. Circle your child's birthday, as well as those of other family members. (3-4) Call attention to his **birth date** when it is no more than 2 months away. "Your birthday will be after Halloween." This gives a timeline or a point for forming a comparison. His/Her birthday now becomes more than just a date in isolation. "

- 3. Teach your child his **birth date**-January 23rd. (3-6)

- 4. Mark special days-a trip, visiting guests, a special event. (3½-6)

- 5. To show the passage of time, show your child his/her baby pictures, as well as pictures of other members of the family. (2-6)

- 6. Teach the **days of the week**. (3-6)
Song- **Tune- 10 Little Indians**
First comes **Sunday**
Then comes **Monday**.
Then comes **Tuesday**.
Then comes **Wednesday**.
Then comes **Thursday**.
Then comes **Friday**.
Then comes **Saturday**.

- 7. What day comes **before**? (4-6)
   What day comes **after**?
   What day comes in **between**?

- 8. (2-3)**Tomorrow**, we will go to the store to buy new shoes."
✓ (3-6) " **Tomorrow, is Wednesday**, we will go to the store to buy new
✓ Shoes."

✓ (2-3) "**Today**, we will go to the store to buy new shoes."
✓ (3-6) "**Today is Wednesday**; we will go to the store to buy new shoes.

✓ (2-3) "**Tomorrow**, you may wear your new shoes to school."
✓ (3-6) "**Tomorrow is Thursday**; you may wear your new shoes to school.

- 9. Learning the **months of the year** is more difficult than learning the names of the week. Young children are more **interested in what is going on now**, as opposed to what will happen in 5 months. Don't refer to dates too far away. If you tell your child that he/she is going on a trip that is 7 **months** away, to the child it might as well be 7 **years** away. (4-6)

Volume and Capacity Measurements-(2-6)
Parent Background
Volume and **capacity** is the amount that a container will hold.

- When asked to choose which container will hold more, most young children will select the tallest.
- Only by comparing the amounts that 2 different containers can hold will the child begin to understand the properties of **volume** and **capacity**.
- We want him/her to discover that the shape of a container does not always determine which one will hold more. "Looks are often deceiving". Give your child time to make these discoveries.
- Even without your intervention your child will develop and refine his/her volume and capacity skills. He/she will make discoveries as he/she:
  - -Stuffs things into the toy box.
  - -Tries to put on clothing that is too small or too large.
  - -Stuffs his/her mouth with food.
  - -Tries to eat more food than his/her stomach will hold.
  - -Overflows the sink.
  - -Tries to squeeze a toy into a hole.
  - -Tries to pour milk into a container.

Vocabulary

• Compare	• Liquid
• More	• Pour
• Less	• Size
• Same	• Empty
• Equal	• Full
• Which will hold more/less?	• Fill
• Not as much	• Almost full

Sample Vocabulary Development
✓ This bottle is "**fill**."
✓ "Pour the water out of the bottle, now it is **empty**".
✓ "Fill your bottle so that it will have **more** water than my bottle."
✓ "**Compare** 2 bottles containing a substance to find the one that has **more** (less)."

Suggested Volume Activities (2-6)
- 1. Give your child a small suitcase, an old plastic lunch box, plastic shoebox, purse, backpack, anything that can be use for carrying things. On their

own, they will discover what will and will not fit into the container. (2-6)

- 2. **Bath time is perfect for exploring volume.** ($1\frac{1}{2}$-6) Even the youngest child will have fun as he/she plays and learns. Provide empty **clear plastic** bottles of different shapes and sizes. Find 2 containers that will hold the same amount, but have different shapes. Provide flat containers, funnels, plastic spoons, different size water bottles, cups, and plastic glasses, anything that will hold water. Put the items in the bathtub and let your child explore. ($3\frac{1}{2}$-6) After you have given him/her **lots of time to explore**, ask simple questions. Which container will hold **more? Same? Equal? Less?**

- 3. Fill a bucket, plastic tub, or large pot with rice, sand, water, or beans. Give your child a measuring cup. Let him/her fill the cup and see how many cups it takes to fill a bowl, coffee mug, and juice cup. (3-6)

- 4. Provide a bucket with water and a measuring cup to water plants. (This is a good outside activity.) Use one cup of water for small plants, two cups for a medium plants, 3 cups for large plants. This activity teaches categorizing skills (according to size), as well as measuring skills. (2-6)

- 4. **Exploring Volume in the kitchen**
 Bake/cook with your child! Read the recipe together and let your child **measure** the ingredients.
   Crispy Rice Treats
      5 c. crispy rice cereal
      5 Tbs. Butter/margarine
      45 large marshmallows
      Melt the butter and marshmallows over low heat. Remove from heat and add crispy rice cereal. Spoon mixture into a greased 9x13 inch pan. Cool and cut into squares. Enjoy!

- 5. Let your child measure sugar and cream for your coffee/tea. ($3\frac{1}{2}$-6)

- 6. Measure milk and cereal for breakfast. (3-6)

- 7. Measure popcorn before and after popping. (3-6)

- 8. When cooking, if it calls for 1 cup of milk, explore combinations that will equal 1 cup. (4-6)
✓ Four - $\frac{1}{4}$= 1 cup
✓ Three- 1/3 = 1 cup
✓ Two- $\frac{1}{2}$= 1 cup
✓ 16 Tablespoons = 1 cup

- 9. **Suggested recipes for measuring:**

- ✓ -Jell-O
- ✓ -Pudding
- ✓ -Chocolate milk
- ✓ -Pancakes
- ✓ -Bread
- ✓ -Cup cakes

- 10. Before cooking measure: (3-6)
- ✓ Rice
- ✓ Pasta
- ✓ Beans
- ✓ Juice
- ✓ Water
- ✓ Ingredients for salad

- 11. At mealtime let your child serve his/her own plate. Use standard measuring tools. "Measure 1 cup of salad for your father." (3-6)
- ✓ $\frac{1}{2}$ cup of salad for yourself
- ✓ $\frac{1}{4}$ c. of rice
- ✓ $\frac{1}{2}$ c. of ice cream
- ✓ 2 teaspoons of gravy
- ✓ 2 tablespoons of peas
- ✓ $\frac{1}{2}$ c. of milk

- 12. Use a teaspoon or a tablespoon to measure and pour liquid dish detergent into the dishwasher. (3-6)

- 13. Measure detergent for laundry. (3-6)

- 14. Select different size boxes. Can you find things that will fit into each of the boxes? Find something that will not fit. (3-6)

Temperature(3-6)
Objectives
- 1. To learn related vocabulary
- 2. To identify the thermometer as a measuring device
- 3. To observe changes in temperature
- 4. To experience temperature

Suggested Temperature Activities (3-6)
- 1. (2-3) Discuss what you will wear today, because it is cold or hot outside.

- 2. Use appropriate vocabulary- warm, hot, cold, cool. (2-6)

- 3. (4-6) Purchase a large outdoor/indoor thermometer and place it insight. For this activity you will need 2 different color washable markers. Use 1 color to mark the afternoon temperature level, (the afternoon temperature is usually the warmest), and the other to indicate the evening temperature. **Compare your findings**. What can you tell me about where the temperature line is in the afternoon? Where is it in the evening? Use descriptive words such as, (up for hot, warm), (down for cold, cool). "Today will be a hot day. Will the mark be up or down?"

- 4. (4-6) Add numbers to compare and contrast temperature readings. The **weatherman predicts** that the temperature will be 80 degrees today? Will the mercury be up or down? This is an on going project. As your child's math and science skills develop he/she will be able to make more connections between the sun and the part that it plays on **animals**, **people**, **plants**, **weather**, and **seasonal changes**. (5-8)

- 5. Use an indoor thermometer to keep track of the temperature. (2-6)

- 6. Measure the temperature up near the ceiling and down close to the floor. Compare your findings. (4-5)

- 7. Find other types of thermometers and discover how they are used. Look around the house for things that use a thermometer. (2-6)

- 8. Take your temperature, using a thermometer. Use a manual thermometer not a digital. (2-6)

- 9. Check the temperature in the **oven**, **refrigerator**, **furnace**, **air conditioners**, **grills**, etc. (2-6)

- 10. Use a candy thermometer to make candy. (2-6)

Linear Measurement (2-6)
Parent Background
- **Linear measurement** is the measurement of length or width of an object that adults compute as **inches**, **yards**, **meters**, **feet**, and **miles** etc.
- (2-4) With young children never think **inches** or **feet**; think **non-standard units**. Explore linear measurement using such things as shoes, blocks, crayons, pencils, paper clips, legos, toothpicks, and straws as measuring units. Use anything that's in your child's world.

You are helping the child to understand the **linear** measurement concept that:
  ➢ the information and it can be helpful

> ➤ it shows how long or short
> ➤ each smaller unit is equal in value
> ➤ it can be used for comparison
> ➤ you apply this information in a certain way
> ➤ when measuring something; you must have enough of one unit to lay each unit end-to-end the full length of the object being measured.

Materials:
Make a set of, "**Happy Hands and Feet**," to be used as rulers for measuring. Trace your child's hand and foot. Cut out at least 15 copies or more of each. To make the copies last longer, make them on cardboard. On one side make happy faces and use them for measuring. Make a second set on a different color paper. These can be used for measuring and comparing 2 different objects.

Use to show a measurement by placing the feet or hands end to end

- 1. Let's measure the bed. We will use your happy feet. (2-6)

- 2. Can you find something that is **taller** than you? (2-6)

- 3. Can you find something that is **shorter** than you? (2-6)

- 4. Use Happy Hands and Feet to measure things around the house. ($3\frac{1}{2}$-6)

- 5. Measure a sidewalk square. Walk heel-to-toe, as you walk, count **how many** of your feet it will take to measure the square. ($3\frac{1}{2}$ 6)
- 6. Measure the length of the sofa, a chair; the distance between two things, a bed, any line. ($3\frac{1}{2}$-6)

- 7. Compare 2 things by measuring one with blue blocks and the other with red. Compare the two by putting the red blocks in a line, end to end. Put the blue blocks next to the red. Which one is **longer, longest, shorter, shortest, more, less**? Match the blocks 1-to-1 to find out which has more? ($3\frac{1}{2}$-6)

- 8. Crate a **growth chart** and attach it to the wall. Every few months measure your child's height. Compare the result. (2-6)

- 9. Trace the outline of your child's favorite toy... a doll or truck, to use for measuring. ($3\frac{1}{2}$-6) **Make several sets.**

- 10. Trace and measure your hand using small paper clips. Compare your hand to the hands of other family members. ($3\frac{1}{2}$-6)

- 11. Use such nonstandard units as (ex. Plastic knives, strips of paper, straws, blocks...) to measure the length of your child's room. (4-6)

- 12. Let your child guess how much string it will take to go around different parts of his/her body. (Guess, measure)

- 13. (4-6) Explore measuring- table legs, cabinets, mirrors, fences, poles, doors, windows, pictures, family members, cat, dog, tree trunks, anything.

- 14. **How far** can you throw a ball? (2-6)

- 15. When measuring small things, try using crayons, blocks, marshmallows, toothpicks, **anything**. Measure plates, forks, napkins etc. (2-6)

- 16. Collect 1 shoe from each family member; put the shoes in order from **largest** to **smallest**. (2-6)

- 17. Cut 1-inch squares from colored paper and use for measuring. Draw long lines on a piece of paper. Glue the squares on the line to measure. Count the squares. (2-6)

- 18. Use "Happy feet," to measure things **vertically**. (3-6)

- 19. Measure your stuffed animals to discover which one is the tallest. (2-6)

- 20. Use the, "Happy Feet," to measure your child's height. Measure and compare each member of your family. (2-6)

- 21. For further exploration, provide a safe **retractable ruler, 12-inch ruler,** a **cloth measuring tape,** and a **yardstick**. Your child will enjoy using each of these standard-measuring tools to explore and make his/her own discoveries and comparisons. (2-6)

Mass Measurement
Parent Background (2-6)
- Mass is measured in **ounces**, **pounds** and **tons**.
- **Don't expect young children to be able to put a number value on the weight of an item with accuracy.** This skill will come in time.
- Like other standard measuring tools scales are based on the abstract understanding of the values of an **ounce**, **pound** and **ton**.
- **Scales** are used to **measure mass**.
- When an adult hears that a box weighs 50 pounds, his/her comparison base gives him/her an idea of approximately how heavy the container feels if lifted.

- At a very early age, children see scales being used. The child can tell you how much he/she weights, and not have a clear understanding of weight.
- Begin the concept of mass measuring by helping your child to understand the concept of **heavy** and **light**.
- Let your child **compare** two things to determine which is **heavier**. From there move to finding things that are **lighter**.

For exploration provide-
- A simple balancing scale
- A small kitchen scale
- A bathroom scale

Vocabulary
- Balance
- Weight
- Weigh
- Compare
- Heavy-heavier- heaviest
- Light- lighter- lightest
- Less
- More
- Equal
- Same

Sample Vocabulary Usage
"This is too **heavy** for you to carry."
"This is not too **heavy** for you to carry, it is **light**."
"This bag weighs more."
"Let's take something out of the bag so that it will **weigh less**."

- 1. A seesaw is perfect for exploring **heavy** and **heavier**. Have the child to sit on the seesaw. What do you think will happen when I sit on the other side? Why? (3-6)

- 2. Hold two things in your hand to determine which is heavier. Find things that are heavier or lighter than the heaviest item. (3-6)

- 3. Brainstorm things that are heavier than a can of soda. Find things around the house that are **lighter** than the can of soda. (3-6)

- 4. As your child's comparison skills develops; **make** the comparisons closer in weight.

- 5. Use two containers; put rice into 1 container. Let your child add enough rice to the second container to make it **heavier** than the first. Make the container **lighter** than the first.
  (4-6)

- 6. Use a small food scale to explore to compare the weights of small items (4-6)

- 7. Provide a bathroom scale for exploration. (4-6)

- 8. Buy a simple balance scale to compare the weight of different items.

$$ Money $$
Parent Background (3-6)
- The concept of money is **based on values** that have been given to **coins, paper currency**, and **rates of exchange** for goods and services.
- Money is a language in itself.
- To use money one must know the value system and how to apply its exchange rates.
- This requires a child to be able to think **abstractly**.
- When adults hold a dollar they abstractly (mentally) know that a dollar is equal to a variety of different coin combinations. These combinations could include:
- ✓ 100 pennies
- ✓ 4 quarters
- ✓ 20 nickels
- ✓ 10 dimes
- ✓ 2 quarters, 5 dimes
- ✓ 2 half dollars
- ✓ 50 pennies, 1 quarter, 2 dimes, 5 pennies
- ✓ 2 dimes, 6 nickels, 2 quarters

There are many other combinations of coins to equal $1.00.

- When dealing with money you are expected to make the exchange rate values in your head and on the spot.
- It is the abstractness of money that makes it difficult for young children to master money skills.
- Abstract thinking and reasoning are **cognitive skills** (thinking) that require a prior knowledge base from which to formulate an answer.
- Imagine yourself in a foreign country, and you are not acquainted with the currency exchange rate. Someone offers you 15 of one coin or 3 of another. Which one do you choose? Think of your frustration. This is just how your child feels.
- Young children equate many coins as more in value. If shown 3 nickels and 1 quarter he/she will probably choose the 3 nickels as being more in value.
- Understanding and applying values to coins takes time and should not begin

Mary Taylor Overton

until the age of 6.
- The American money system is based on values of 10's.
- Counting money requires that your child be able to count by 1's, 5's, and 10's.
- He/she must also recognize and know the value of each coin.
- When introducing money (2-5) **focus** on recognition of the **coins** rather than their **values**. **Do not begin with the exchange rate.**
- Start with the **exploration** of coins. The penny comes first because its color makes it stand out from the other coins. It is best to use **REAL coins** and not plastic or cardboard fake ones!

Objectives
1. To learn related vocabulary
2. To explore coins
3. To identify coins
4. To learn the value of a pennies, nickels and dimes
5. To understand that more expensive things cost more money

Vocabulary
-
Money
- Coins
- Dollars
- Penny
- Nickel
- Quarter
- Half dollar

Suggested Money Activities (3-6)
Using a **magnifying glass**, have your child study a penny. Ask, "What do you notice?" He/she may answer, "A man with a beard." You can tell him/her that Abraham Lincoln lived a long time ago, before there were cars! And he helped make rules for our country and was important. (Don't get into a huge history lesson, unless he/she is really curious and continues asking you questions.) Flip the coin over. Ask, "What do you notice on this side?" (If you look very carefully on a newer penny, you can see the statue of Abraham Lincoln in the middle of the Lincoln Memorial) Continue discussing, "What color is this penny?" "Is it heavy?" "Does it have words on it?" "Numbers?" "What shape is it?"

- 1. Hide **pennies** around the house. Search for the treasure of pennies. Collect and count.

- 2. Hide **pennies** in a sandbox. Dig for pennies with a shovel.

- 3. Flip pennies. Play **Heads** or **Tails!**

- 4. How many pennies can you pickup at one time?

- 5. Make outlines or designs using pennies.

- 6. **Stack** pennies.

- 7. **Roll pennies** and take them to the bank.

- 8. Collect pennies. Compare older pennies to newer ones. Ask, "Why do you think this one is not as shiny as this one?"

- 9. Locate a **special penny** with his/her **birth year**. Help your child keep it in a special place. "This penny is as old as you!"

- 10. Play store, using pennies for payment.

- 11. Collect pennies and use them for measuring small things.

- 12. Next introduce a **nickel**. Help your child use a magnifying glass to study its features. Ask, "What do you notice?" He/she may say, "A man with a ponytail." Let him/her know that his/her name was Thomas Jefferson and just like Abraham Lincoln from the penny; he lived long ago and helped make rules for our country. (This is a simple explanation that young children can understand because by this age they understand about rules.) Study the flip side of the nickel; asking questions like you did for the penny.

- 13. With your magnifying glass compare the nickel to the penny. Contrast. Discuss. They are similar because both have men's faces on the front and buildings on the back. They both have numbers and letters on them. Both are round. They are different in their color and size. The men's faces are looking in opposite directions. Jefferson has a ponytail in his hair and Lincoln has a beard. The building on the back of the nickel has a dome on top and the building on the penny does not.

- 14. New nickels have recently been issued that have a different back. Collect a small cup of nickels, making sure that some of the newer nickels are included. Sort the nickels by matching the backs.

- 15. Hide nickels in a sandbox. Follow directions for penny games only using nickels instead!

- 16. Explain that a penny is worth 1 cent, while a nickel is worth 5 cents.

Mary Taylor Overton

Give him/her a pile of pennies. Have him/her count groups of five and trade each with you for a nickel. (5-6)

- 17. Count by 5's using nickels (5-6)

- 18. Introduce the **dime** by studying it under a magnifying glass, similar to the penny and nickel. Guide your child to make his/her own discoveries.

- 19. Compare the dime to the penny and nickel. Contrast. They are similar because they are round, have a man's face on each, and have numbers and letters written on them. They are different in size, color (the copper penny vs. the silver dime and nickel), the back (dime has a torch and branches-no building), and the edge of the dime is rough- not smooth like the nickel and penny.

- 20. Play games with dimes similar to the ones described for the penny. Hide, flip, roll, and count the coins. (3-6)
- 21. If your child is ready, discuss value. Count groups of ten pennies and trade for one dime. If you feel your child is comfortable with this concept, you may go on to explain that 2 nickels equal 1 dime. (age 6) Trade 2 nickels in for 1 dime. Please don't move too quickly. If he/she seems confused or uninterested do not continue activity. Revisit when the child is older. (5½-6)

- 22. Count by 10's using dimes. (4-5)

- 23. Introduce the **quarter**. Study it under a magnifying glass and discuss. Again follow instructions for the penny introduction. Make a special note to explain to your child that while the front of each quarter looks the same with George Washington's face on it, the back of the quarter may have one of many designs on it-(due to the introduction of state quarters). (4-6)

- 24. Play a quarter games similar to penny games.

- 25. Compare and contrast to other coins. Help your child to notice that quarter is larger.

- 26. If your child is ready, encourage him/her to count 25 pennies to trade in for each quarter and discuss value. (5-6)

- 27. After all of the coins have been introduced, sort, graph, and count a jar of mixed coins. (4-6)

- 28. Using 2 **quarters**, 2 **dimes**, 2 **nickels**, and 2 **Pennies** play memory. Lay the coins on the floor face up, and cover them with index cards. Flip two cards at a time and try to make a match. Take turns. Increase difficulty by increasing numbers of coins (ex. Use 4 of each instead of 2). Also, lay the coins face down to increase difficulty!

- 29. Play store. Label items 1 dime, 3 pennies, etc. Or, if your child is aware of the value, items can be marked 25 cents, 1 cent, 5 cents, or 10 cents. (4-6)

- 30. Set-up a lemonade stand. Lemonade could cost a dime and a cookie a nickel. (4-6)
- 31. Use coins for game pieces. Each player chooses a different coin to be his game piece. (2-6)

- 32. Play in the bathtub with coins. Which sinks faster? (3-6)

- 33. Drop coins on carpet. Which bounces farther? (3-6)

- 35. Visit the gumball machines. Use correct coins. (4-6)

- 36. Make designs using coins.

- 37. Build towers or columns with coins. Which coin can you use to make the tallest column?

- 38. Collect and save a shiny penny, nickel, dime, and quarter that were minted in your child's birth year.

- 39. Guess how many quarters it will take to measure across a plate, place mat, a napkin, etc. Try this with other coins too. Guess, measure, and count.

- 40. Use a ruler to indicate the 1-inch mark. How many quarters or other coins will it take to stack to the 1-inch mark? Guess and stack. (5-6)

- 41. How many coins can you stack without them toppling?

- 42. Sort coins and put them into coin sleeves to take to the bank.

- 43. Will it cost more money to buy a car or a cake?

Problem Solving
Parent Background (2-6)
Problem solving skills are **organizational** and **management** tools. These skills include **estimating**, **sorting**, **recognizing patterns**, and **graphing** to name just a few. These skills are a way of **gathering** and **organizing** information so that the information can be evaluated. Problem solving skills allows the child

to use his/her prior knowledge about a topic to formulate strategies for understanding new information.

Estimating-A Problem Solving Skill
Parent Background (4-6)
- Estimation is a life experience skill that crosses every aspect of our lives.
- It is simply, **your best guess**.
- **Estimation is not just a wild guess**, it is a guess that is based on our **prior knowledge** and **comparison experiences**.
- When an adult looks at a person, he/she uses prior knowledge to estimate the person's age, height, and weight. Based on this knowledge base, he/she can mentally estimate the person's height to be approximately 6 feet tall, age-30ish and weight, about 180 lbs.
- We use estimating skills everyday without giving it a second thought. You estimate time, numbers, weight, height, length, money, size, how much, how many etc.
- You estimate how long it will take you to do a task? How many pieces of chicken will fit into the pan? The amount of food required to feed 10 guests? If there is enough room to fit into a parking space? "There were about 150 people at the wedding." "The groceries in my cart will cost about $35.00." "This dress looks to be my size."
- At a very early age your child has also used his/her estimating skills. Children estimate the distance of a toy, "If I tiptoe, I will be able to reach it?" You have heard a child say, "I ate 900 pieces of candy." **Only experience and time will improve his/her accuracy.**
- Begin by helping your child to develop **comparison** skills. Ask such questions as, "Which is this **taller** or **shorter**? Will this fit in here? Which is the **biggest**?"

Objectives-
1. To build a prior knowledge base
2. To understand what a guess is
3. To understand when it is appropriate to guess
4. To understand that after making an estimation, you need to check your answer to see how close you are to the actual amount

Vocabulary
- Guess
- Count
- Check answer

Sample Vocabulary Development
✓ "I am going to **guess** that my foot can fit into the shoe."
✓ "Can you guess which shoe is too large?"

Suggested Estimation Activities (4-6)

- 1.   Are you **taller** than the table?

- 2.   Which is **longer**?

- 3.   Which one is **larger**?

- 4.   Will all of these toys **fit** into the box?

- 5.   Can you make this building **taller** than you are?

Make your problems age-appropriate.

Always guess and then **check your answer** to see how close you are. This gives instant feed back and positive re-enforcement. It also gives your child an opportunity to refine his/her estimating skills. If the guess is too high, perhaps next time it will be lower. (4-6)

Keep the quantity numbers low and the size of the objects large.

- **Make an Estimation Jar.** (4-6) Choose a large clear plastic jar. Start by putting 5 or 6 items into the jar. Your child is to guess how many items are in the jar. Increase the number of items as his/her guesses are more accurate and number recognition skills increase.

    Suggested Things to Estimate(4-6)

Never pack the estimating jar. Your child needs to be able to see the individual objects.

<u>(Large jar)</u>   <u>(Small jar)</u>

- Candles
- Forks
- Spoons
- Eggs
- Medicine bottles
- Golf balls
- Tennis balls
- Large pasta
- Large legos
- Clothes pins
- Straws
- Rolled socks
- Toothbrushes
- Crayons
- Pencils
- Pens
- Markers
- Mushrooms
- Hair rollers
- Combs
- CD cases
- Pretzel rods
- Small blocks
- Dog biscuits
- Lemons/ limes

Jelly beans
Birthday candles
Large buttons
Earrings
Medicine bottle tops
Small pasta
Large paper chips
Tooth picks
Hair clips
Beads
Quarters
Marker tops
Pen tops
Pebbles
Poker chips
Grapes
Whole nuts in a shell

- 7.   Use **seasonal candies** for **estimating**. Put 8 jellybeans in the jar.
- -Halloween- Candy pumpkins
- -Thanksgiving- Corn candy
- -Christmas- Candy canes
- -Hanukah- Golden Gilt
- -Valentine's Day- Candy hearts

(Not only are they fun to work with, but also fun to eat)

- 8. **How many** books are in a stack?

- 9. **Guess** how many shoes are in a pile?

- 10. Show me **how much** string will it take to go around this ball?

- 11. Take advantage of any opportunity to use **estimating**, they can occur at anytime.
- 12. How many small blocks will it take to **measure** a line?

- 13. How many cups of water will it take to fill this glass?

- 14. Guess how many steps it will take to walk to the car?

- 15. How many dog biscuits will fit into the container?

- 16. **Estimate** the number of large pieces of pasta on your plate? **Count** as you eat.

- 17. How many onions are in the bag?

- 18. How many potatoes are in a 5-lb. bag?

- 19. How many green peas will fit on a spoon?

- 20. How many fruit snacks are in a small pouch?

- 21. Estimate the number of words in a sentence?

- 22. Estimate how many steps it will take to walk from the sofa to the kitchen?

- 23. Estimate the number of plastic figures you can pick up with one hand?

- 24. How many toys can you pick up before I count to 50?

- 25. How many jellybeans can you pick-up-using one hand? 2 hands?

Patterning-A Problem Solving Skill (2-6)

Parent Background

- Recognizing patterns teaches the child how to look for **logical relationships**. Both science and math are based on recognizing patterns.
- Developing patterning skills should begin at a very early age ($1\frac{1}{2}$-6). Start with a series of simple rhythm clapping games.
  Example: -clap (stop) clap (stop) clap (stop.)
- When teaching patterns you must perform the **rhythm** of the pattern, and verbalize the pattern at the same time. Children need to hear and perform the pattern at the same time.
- A pause between each set will help to create a strong rhythmic beat to the pattern. This also makes it easier to distinguish a beginning and an end to each pattern's segment.
- The skill of recognizing patterns requires time. Looking for patterns is a life long open-ended skill. New patterns will emerge throughout ones life.

Objectives

Mary Taylor Overton

1. To learn related vocabulary
2. To learn to recognize a pattern
3. To discover where patterns can be found
4. To extend a pattern
5. To create a pattern

Vocabulary
- Patterns
- Repeat
- Copy my pattern
- Make your own pattern
- Extend the pattern (continue)
- Can you read the pattern?

Sample Vocabulary Development
✓ " Look, I see a **pattern**."
✓ "I am going to read the pattern."
✓ "The pattern is....."
✓ "Can you read the pattern?"

Suggested Pattern Activities-
- 1. **Begin with simple patterns** (2-6)
✓ -Clap, jump/ Clap, jump/ clap, jump
✓ -Clap, stomp/ clap, stomp/ clap, stomp
    (Make up your own patterns)

- 2. **Move to a more complicated pattern**
 ✓-Clap, clap/ jump, jump/ clap, clap/ jump, jump

 ✓ -Stomp, stomp/ clap, clap/ stomp, stomp/ clap, clap

 ✓ -Shake, shake/ jump, jump/ shake, shake/ jump, jump

- 3. **More complicated patterns**
✓ -Clap, clap, clap/ stomp, stomp/ clap, clap, clap

- 4. As your child grasps simple **patterning skills**, add more complicated patterns. Example- **Clap, clap/ stomp, stomp/ jump, jump/ clap, clap**
    Other suggestions-
✓ Snap fingers
✓ Shake your hands
✓ Shake your head
✓ Twist your body
✓ Bend your knees
✓ Tap your toes

- 5. (4½-6) Cheer, stomp, jump,
  cheer, stomp, jump
  kick, clap, stomp
  kick, clap, stomp

(Remember to Say the Pattern Out Loud) (2-6)

- 6. Patterning activities for young children should always be **concrete**, **hands-on**, **visual**, and **vocal**. Create a pattern and have your child continue the pattern.

- 7. Use materials such as blocks of 2 different colors, blue and red. Display 1 red and 1 blue. Keep repeating this pattern. Verbalize the pattern as you construct it. Have your child say the pattern.
  (3-6) This pattern will be (red-blue) (red-blue).
  Let your child choose 2 things to use for the pattern.

- 8. Still using **two** colors, move on to a more complicated pattern-(Red, red, blue/ red, red, blue) (2-6)

- 9. As your child progresses add a **third** color. A suggested pattern might be (car, sun, bed) (car, sun, bed) or (red, blue, green) (red, blue, green) (3-6)

- 10. Encourage the child to create patterns. Try not to make the pattern too complicated. (2-6)

- 11. Find patterns anywhere-on clothing, dishes, walls, in nature-etc. (2-6)

Sorting-A Problem Solving Skill
Parent Background-(2-6)

- **Sorting** is another organizational skill.
- It is a way of grouping information.
- Things are sorted into groups according to a common feature.
- Adults and children sort without realizing it. When in the grocery store, we sort through the strawberries to find the firm ripe ones. If you like green jellybeans, you sort to find the green ones.
- Things can be sorted in many ways-by color, shape, size, age, membership, use, ownership, material make-up etc.
- Sorting is a natural skill for children. There will always be opportunities to refine sorting skills.

Suggested Sorting Activities-(2-6)

- 1. Make a **sorting box**. Select a large flat container, one that will allow all items to be easily seen. Fill the box with discarded small items-any old game pieces, odds and ends, beads etc. Let your child just play and explore the contents of the box. You will find that as he/she explores, he/she will **sort** on his/her own. Later assign a sorting task by an attribute.

- The older the child, the more attributes he/she will be able to handle.
- **Sample-Vocabulary Development**
✓ "Can you find all of the beads?"
✓ "Now, can you find all of the yellow beads?"
✓ (Work with your child. Sit and ask questions and model the task.)
✓ "Does this one belong here or there?" Why?

List of attributes for sorting-
- Dominoes-number of dots
- Poker or game chips-color-size
- A bag of buttons by:
✓
Color
✓ Size
✓ Shape
✓ Number of holes
✓ Shiny
✓ Dull
✓ Metal
- Favorite
Beads-color-size-shape
- Beans (dried)-color-size
- Candy- shape-color-size
- Jelly beans-color-size
- M & M's-color
- Pasta- shape
- Plastic figures- job-color

- 2. The laundry room is a good place for sorting. Put clothing into piles according to color, ownership, sleep clothing, socks, jeans, and shirts. Think of other ways to sort. (2-6)

- 3. Load or unload the dishwasher by sorting forks, knives, spoons, bowls etc. (2-6)

- 4. Sort a deck of playing cards by number, color, or picture. (2-6)

- 5. ($2\frac{1}{2}$-6) **Bean Soup Game**- This game uses imaginative play, which young children love. If you are relaxed and having fun with your child, he/she will not even realize that he/she is doing a math lesson!! You will need a bag of 15 bean soup beans. These beans can be found in the dried bean section of your supermarket. Give your child a bowl/or bowls to put selected kinds of bean to make soup. Put all of the red beans in this bowl. Make-up another game using the same package of beans. Sort by color, size, kind, and shape.

- 6. Sort tri-colored pasta before cooking. (2-6)

- 7. Make a necklace of multi-colored beads- String by a specific color, size, or shape. (3-6)

- 8. Small cars-Use a shoebox as a garage. Sort by colors, kinds, or use. (2-6)

- 9. **Animals**- Plastic animals may be purchased by the bag from a dollar store. Sort by-number of legs, covering-feathers, fur, zoo, and farm animals. (2-6)

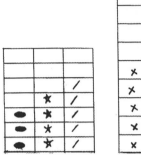

- 10. **Make a Button Box.** Collect old buttons in a box. Sort the buttons by-size, number of holes, shape, or by material. (2-6)

- 11. Put keys on a key ring by color, shape, or by size. (2-6)

- 12. Also try sorting the following: (2-6)
  - ✓ Toys
  - ✓ Shoes (size, color ownership)
  - ✓ Coins
  - ✓ Pieces of kid's cereal
  - ✓ Mail by size, color
  - ✓ Groceries (what goes into refrigerator, cabinet, freezer)
  - ✓ Colored toothpicks
  - ✓ Colored paperclips
  - ✓ Plastic tops
  - ✓ Jewelry
  - ✓ Flowers
  - ✓ Leaves
  - ✓ Buttons
  - ✓ Plastic figures

Graphing-A Problem-Solving Skill (3-6)
- **Parent Background** (3-6)

Graphing? Yes, graphing, it is a wonderful way to organize information.
- Graphing provides an easy to read summary that can be used to compare

and contrast information.
- A graph is a **visual summary picture**.
- Graphing is an extension of sorting.
- It is a way of compiling information in an orderly way.

Objectives-
1. To learn related vocabulary
2. To use graphing as an organizational tool
3. To graph things according to common attributes

Vocabulary
-
Graph
- Sort
- Count
- More
- Less
- Equal
- Match 1-to-1

Procedure-
When **graphing** objects, it is important to line the columns neatly and orderly so that the results may be easily seen.

Sample Vocabulary Development
Steps 1-6 are appropriate ($3\frac{1}{2}$-6)
Start with a stack of red and green objects.
- 1. "**Sort** by putting all of the red in this row, and all of the green in this row."
- 2. **Count** each **row**.
- 3. "Are there **more** red or green?"
- 4. "Let's **match the columns 1-to-1** to find out which has more?"
  Slide your finger across the row to match 1 red to 1 green.
- 5. After matching, your child will be able to see which column has **more** or **less**.
- 6. Let's count to find out **how many more** reds there are"
- (4-6) The concept of **less** is much harder than, **more**. Move on to less, only after your child for has an understanding of **more**.
- 7. Use 2 columns graphing for 3-4 year olds.
- 8. (4-6) Compare 3-5 columns.

Directions for making a reusable graph:
An old plastic place mat may be used for making a small graph (no design or turn over on the wrong side). A larger graph can be made with a one-yard piece of white fabric (cut to 36 x36), or a heavy plastic tablecloth. With a permanent marker make large equal columns. (4-6)

(You may substitute poster board).
Most of the sorting activities in this book can also be used for graphing.

Suggested Graphing Activities(3-6)

- 1. Give your child a bucket of checkers. Have him/her to close eyes and pick out a handful. Put all of the red checkers in a row on one side and black on the other. (3-6)

- 2. Graph coins by size or color. This is an excellent motivation for learning the names of the coins. (3-6)

- 3. Are there more windows or doors in your house? As you go from room to room, use red and blue poker chips- **red** (windows) and **blue** (door), put a chip in the bucket to represent each window and each door. Graph by color to determine if there are more windows or doors in your house? Less? Equal? (4-6) Find other things to graph

- 4. Take a handful of multi-colored kid's cereal and graph. (3-6)

- 5. Think of your friends and their family members; use two different colored poker chips to tally the number of boys and girls. Graph to find out if there are more girls or more boys? How many more? Less? Equal? (4-6)
- 6. Call grandparents and friends to collect data. **What is your favorite color?** (Give 2 choices depending on the skill level) Graph your findings.

✓
Color
- ✓ Fruit
- ✓ Vegetable
- ✓ Flavor of ice cream
- ✓ Book
- ✓ Food
- ✓ Sport
- ✓ Cake

- ✓ Flower
- ✓ Animal
- ✓ City
- ✓ Game
- ✓ Holiday
- ✓ Video
- ✓ TV program

Mary Taylor Overton

Understanding Your Body
Parent Background (2-6)

- The human body is made up of systems.
- The **brain** is the body's computer.
- The brain gathers, stores information, coordinates, and interconnects all of the bodies systems.
- Our bodies are covered inside and out with **sensory receptors**.
- The brain depends upon the sensory receptors located on the body's **5 senses** to gather information.
- The 5 senses: **ears-hearing, eyes-seeing, skin-touching, nose-smelling and tongue-tasting**, work together as they control every aspect of our daily lives. All **five senses** are interconnected and all have major impacts our quality of life.
- The **brain** is the body's problem solver. It is constantly **retrieving** stored information to understand and make connections to new discoveries.
- When a receptor on the body receives a **message**, it is immediately sent to the brain. The brain uses information from prior experiences to process and identify messages.
- Based on the connections made from prior stored information the brain plans a reaction.
- Humans as well as animals depend upon their senses for their daily survival.

Your Body Objectives- (2-6)
1. To identify and name body parts
2. To use related vocabulary
3. To observe the uniqueness of each of the body's systems
4. To identify our 5 senses
5. To understand how the body's systems are interconnected
6. To understand how people and animals depend upon their senses for their daily survival

Vocabulary-
- 5 Senses-
✓ ear-hearing-
✓ eyes- seeing
✓ skin-touching
✓ nose-smelling
✓ tongue- tasting
- Brain- The body's computer
- Heart- Beats, pumps blood,
- Skeleton- The body's support system-bones, muscles, movement
- Lungs- Breathe, air, oxygen, inhale, exhale
- Energy- Power

Our Senses (2-6)
Parent Background

Hearing-Sense of Hearing-Ears

The shape and the makeup of the ear are designed for catching and funneling sounds into the ear. The ear works together with the brain in identifying sounds. The eyes also help the ears to work.

Suggested Hearing Activities (2-6)

- 1. Identify parts of the ear.

- 2. Close your eyes and try to identify sounds. (2-6)

- 3. To hear an **echo**, yell into a large pot. Experiment with different size pots. Try using cardboard boxes. (2-6) Discover what is causing the echo. (4-6)

- 4. Take a "**Listening Walk.**" (Walks may also be taken for each of the other 4 senses) (2-6)

- 5. To feel **vibration**, hold your hands against your cheeks as you hum a tune. Make long, short, high, and low notes. (3-6)

- 6. Make a vibrating instrument by stretching rubber bands over an empty tissue box. For varying vibrations use rubber bands of different thicknesses. Try inserting a pencil (between the box and the bands) under one end of the bands. Pluck the bands and listen to the **vibrating sound.** (4-6)

- 7. Examine a guitar; strum it to look and feel the vibration of the strings. Pluck each string to find out what happens. (3½-6)
- 8. Make a shaker drum. Glue the outer edges of 2 paper plates face to face. Before sealing, add a few dried beans. Decorate and add colorful streamers. (2-6)

- 9. Use a metal or wooden spoons to drum on metal pots, empty glasses, empty cans, plates etc. (2-6) Is there a difference? If so why? (5-6)

- 10. Brainstorm things that are: **loud**, **noisy**, **soft,** or **quiet** (2-6)

- 11. Discover how **hearing aides** help people to hear. (4-6)

- 12. Discuss the importance of not listening to music that is too loud. (2-6)

- 13. Make a **megaphone** from a cardboard paper party hat. Cut off the pointed top to create a mouthpiece. Explain that electric megaphones are called **bullhorns**. (2-6)

- 14. **Mammals** have outer ears. Look at pictures of mammals to compare the size, placement, and shape of their ears. Compare your ears to those of a monkey or other mammals. (lions, elephant, or a rabbit) (2-6)

- 15. Owl's ears are **asymmetrically** placed on their skulls, allowing them to hunt without seeing its prey. Many owls have **tufts** on their heads that appear to be ears, but they are actually feathers. Learn more about owls. (2-6)

- 16. Whales and many insects hunt by **echolocation**. They rely on sound vibrations bouncing off of objects to locate their food. Learn about echolocation. (4-6)

- 17. Most animals depend upon their sense of hearing for survival. Investigate how animals use their sense of hearing. Example:
  - **Robins** can hear an earthworm moving around underground.
  - **Bats** flying at night use echo sound to locate insects, as both animals are flying through the air.
  - **White-tailed deer** have large ears that turn independent of the other.
  - **Beavers** slap their tails on the ground to warn other beavers of danger.

- 18. Hold a seashell to your ear. What do you hear? (2-6) Why do you hear a sound? (5-6)

- 19. Create clapping or stamping patterns. (1-6)

- 20. Use **earphones** attached to a recorder or radio to enjoy a story. (2-6)

- 21. Make a horn from a paper towel roll or a drum from an empty coffee can, oatmeal box etc. Use the instruments to have a parade. (2-6)

- 22. On a warm summer's night listen to the sounds of crickets chirping or frogs crocking. Listen for other **night sounds**. (2-6)

- 23. Fill 5 or 6 glasses with graduated water levels. Tap the glasses with a metal spoon and listen to the differences in sound. (4-6)

- 24. Use your voice to make **high** and **low pitch** sounds. (2-6)

- 25. **Sing** the musical scale. Use an instrument to play the scale. ($1\frac{1}{2}$-6)

- 27. **Whisper.** ($1\frac{1}{2}$-6)

- 28. Hide a timer; use your ears to find it. ($1\frac{1}{2}$-6)

Seeing- Sense of Seeing Eyes (2-6)
Parent Background
- Everything gives off light waves.
- The light waves travel through the air.
- The eye is the organ that gathers light wave images.
- The images of an object are sent to the brain where they are identified.

Suggested Eye Activities
- 1. Learn the main parts of the eye. (2-6)

- 2. Tears, eyelashes, and lids all have a purpose. Tears lubricate and wash away debris. Eye lashes and lids keep dust from getting into the eye. Take a closer look to discover how they protect your eyes. (3-6)

- 3. Brainstorm things you like to see. (3-6)

- 4. Discover how eyeglasses help people to see. (4-6)

- 5. Visit an eyeglass-making store to find out how glasses are made. (4-6)

- 6. Take an eye examination. (3-6)

- 7. Discover what it means to be **nearsighted** or **farsighted.** (4-6)

- 8. Learn how contact lens helps people to see. (4-6)

- 9. **Optical illusion** (4-5) Look at picture books that are made for finding optical illusion pictures.
- ✓ Look straight ahead. Try to touch your fingertips without looking directly at your fingers.

- 10. Discover what it means to be **colorblind.** Colorblind means that you cannot distinguish the colors yellow or blue. Men are more likely to be colorblind than are women. (4-6)

- 11. Investigate how animals use their sense of seeing. Example- Eagles have good eyesight. (4-6)

- 12. Look at a picture of animal's eyes. Compare the placement of their eyes. Example- owls, horses, crocodiles, robins etc. (2-6)

- 13. Wear **sunglasses**, try glasses with different tints. Discuss why we should not look directly into the sun. (2-6)

- 14. Learn about **Seeing Eye** dogs. (3-6)

- 15. Create a mini **obstacle course**. See how fast you can run the course. (2-6)

- 16. Look through a pair of **3D glasses** (2-6)

- 17. Discover some animals that are blind-**moles** and **earthworms**. (2-6)

- 18. Blindfold your child and give verbal directions to follow. (2-6)

- 19. Look at your **reflection** in a mirror. (2-6)

- 20. Learn the colors of a **rainbow**-red, orange, yellow, green, blue and indigo. Draw a rainbow. (4-6)

- 21. Mix **primary colors** to make **secondary colors**. (4-6)
  ✓ Red and yellow = orange
  ✓ Blue and yellow = green
  ✓ Blue and red = purple

- 22. Look at colors on a **color wheel**.(4-6)

- 23. Colors exist in **shades**. Select a solid blue object; now find an object that is **lighter** and one that is **darker** in color. Try other colors. (4-6) Visit a paint store to get green paint chips in different shades. Get a several different colors or your child's favorite color. Have your child put the colors in order of **light** to **dark**.

- 24. Get samples of paint chips from the hardware store. You will need 2 single chips of the same color. Cut the colors apart and use as a matching game. (3-6)

- 25. Discover why scientists use **microscopes**. Purchase a cheap microscope to make discoveries. ($3\frac{1}{2}$-6)

- 26. Have fun Looking through a **kaleidoscope**. (2-6)

- 27. Look at the night skies through a **telescope**. (2-6)

- 28. Use a **magnifying glass** to view small things. (2-6)

- 29. **Binoculars** allow you to get a closer look at things in the distance. View objects through a pair of binoculars. (2-6)

- 30. Some animals like fish and snakes cannot close their eyes; discover how their eyes are protected. (3-6)

- 31. Insects have **compound eyes**. Compound eyes are made-up of many small lenses. Learn more about compound eyes. (4-6)

- 32. Spiders have 8 eyes. Discover other animals that have more than 2 eyes. (3-6)

- 33. Compare the placement of a frog's eyes to a bird's. (2-6)

- 34. For protection many animals use their skin colors to **camouflage**. Learn about camouflage. (3-6)

- 35. **Nocturnal** animals have eyes that are suited for night hunting. Discover some nocturnal animals. (2-6)

- 36. Use your eyes to find patterns. (2-6)

- 37. **Blink** your eyes fast and slow. (2-6)

- 38. Look at an object and sketch it. (4-6)

- 39. Play a memory matching game. (2-6)

- 40. Try communicating with each others without speaking. (2-6)

- 41. Get a book on **sign language** and learn to sign a few words. (3-6)

- 42. Look through reading glasses. (3-6

- 43. Discuss care of the eyes-Don't look directly into the sun. (3-6)

- 44. Discover why some people go blind. (3-6)

- 45. Sort a handful of colored cereal. (2-6)

- 46. Many birds **migrate**. Some scientists believe that these birds use landmarks to guide their travel. Learn more about migrating birds. (3-6)

- 47. Help your child to discover the properties of **reflection**. (3-6)

- 48. Go on a **Reflection hunt**-mirrors, water, a freshly polished wooden table, a glass of water, storm door, windows, TV screens, CDs, computer screens, glass in picture frames, silverware, faucets, chrome items, a car's body etc. How many other places can you discover? (3-6)

- 49. Provide your child with his/her own **unbreakable** mirror, one that may used at any time. Encourage its use. (2-6)

- 50. Hold objects over tub of with water, to see their reflection in the water. (2-6)

- 51. Use **reflective** surfaces to make funny faces and to reflect bright shinny objects. (2-6)

- 52. Make a simple design on a piece of paper. To create a symmetrical picture, place a mirror in front of the design. What did you discover? 2-6)

- 53. Look at Chinese writing. (2-6)

Touching- Sense of touch- Skin (2-6)
• Parent Background
- The skin is the largest body organ. It covers the whole surface of the body.
- The sense of touch is controlled by sensors located on the skin. These sensors send messages to the brain and the brain reacts. Example: When your hand touches a hot stove, the sensors on the hand immediately send a message to the brain. The brain realizes that the hand is on something hot and in a split second a message is sent to the muscles in the hand to move quickly.

- 1. **Descriptive Words for the Sense of Touch**

Rough, smooth, bumpy, soft, hard, cold, hot, dry, stiff, flimsy, slick, sharp, dull, fluffy, wet, damp, slippery, thick, thin, puffy etc.

- 2. Brainstorm things that you like to **touch**. (3-6)

- 3. Go on a, "**Feeling Hunt**." "You must find 3 things that are smooth."- **Fluffy, smooth**, or **rough** etc. (2-6)

- 4. Feel an object; use at least 2 descriptive words to describe it. Example: Ice is **cold** and **slippery**. (wet, hard, smooth, slick, watery) (2-6)

- 5. Discover how blind people use **Braille**. (4-6) Use a straight pin to make holes in a piece of paper. Turn the paper over and feel the Braille marks.

- 6. Close your eyes and identify different objects. (2-6)

- 7. Brainstorm things that are **sharp** or **dull**. (4-6)

- 8. Feel different grades of sandpaper. Use sandpaper to smooth a rough surface. (2-6)

- 9. **Earthworms** do not have eyes and must depend upon their sense of touch. Learn about these animals. (3-6)

- 10. **Fish** have a **lateral line** that run the length of both sides of the body. These lines are able to sense vibrations through the water. Learn more about fish. (3-6)

- 11. Play a searching game. Each person is to find 3 things in each of the following categories-smooth, bumpy, wet or hard etc. Use the descriptive words. (4-6)

Smelling- Sense of smell. Nose (2-6)
Use the descriptive vocabulary **odor, pleasant, unpleasant** etc.

- 1. Learn the parts of your nose- **nostrils, bridge of your nose.**(2-6)

- 2. Brainstorm things that you like to **smell**; things that you do not like to smell. (3-6)

- 3. Bake cookies and describe the **aroma**. "What is that aroma?" (2-6)

- 4. Collect things to smell. (2-6)

- 5. Take a,"**Smelling Walk.**" Find things that you can smell. (2-6)

- 6. Color with scented markers. (2-6)

- 7. Discover how animals use their sense of smell. ($3\frac{1}{2}$-6)

Mary Taylor Overton

> Elephants can smell ripe fruit miles away.
> Skunks use an unpleasant odor for protection.
> Snakes have an excellent sense of smell.
> Blowflies can smell rotting meat from long distances.

- 8. Many breeds of dogs are trained to identify specific smells. Learn about how these dogs are used. The **bloodhound** is a breed of dogs with a keen sense of smell. They are often used to follow the scent of lost or missing people. Learn to identify this breed of dogs. (2-6)

Tasting sense of taste-tongue
Parent Background (2-6)

- Although, the tongue controls the sense of taste, your sense of smell and sight lend helping hands. Even before the **taste bud receptors** located on the tongue have received a **message** from the food, the receptors in your eyes and nose are already at work.
- The receptors in your eyes have received a visual message and your nose an aromatic message and both have sent their clues to the brain.

- When your mouth and tongue receptors taste the food they too send messages to the brain.
- The brain uses its stored database of information, along with the new messages to evaluate the food and respond to it.

Suggested Tasting Activities (2-6)

- 1. Learn the different parts of the mouth.

- 2. Think of things that you like to eat. (2-6)

- 3. Let your child use a magnifying glass to get a closer look at your tongue and his. (2-6)

- 4. Use your **taste buds** to send messages to your brain. Taste different foods. Try foods that are sour, sweet, salty, and bitter. Try tasting new strange foods. (2-6)

- 5. Hold your nose and taste foods. What did you discover? Try to identify the food. (2-6)

- 6. The elephant's nose is at the end of its trunk. Learn more about how these animals use their trunks. (2-6)

- 7. Many insects taste with their feet. Discover how other animals taste. (2-6)

Your Skeleton (2-6)

Young children are fascinated and mystified by the human skeleton. When they look at their hands or feet they think- **1 big bone**. To give your child an idea of his/her skeleton's structure help him/her to make a skeleton. (a pattern is provided)

Parent Background-

- Animals are classified as either **vertebrates**, those with backbones or **invertebrates**, those without.
- Mammals, birds, reptiles, amphibians, and fish have backbones and are classified as vertebrates.
- Humans are classified as **vertebrates**.
- Our **skeletal** (bones) system works in conjunction with **joints**, **muscles** and **tendons** to give the body shape and movement. They allow us to stand, walk, run, sit, hop, sit, wiggle, bend, and jump.
- The skeleton also provides protection to soft vital organs.
- Humans are born with over **300** bones. As they grow, some of these bones become fused together.
- Adult humans have **206** bones.
- Humans also have excellent use of their opposable thumbs. Their thumbs allow humans to accomplish complicated task using their hands.
- They also have a large toe placement that gives balance and the ability to walk upright on 2 feet.

Reference for Making a Skeleton

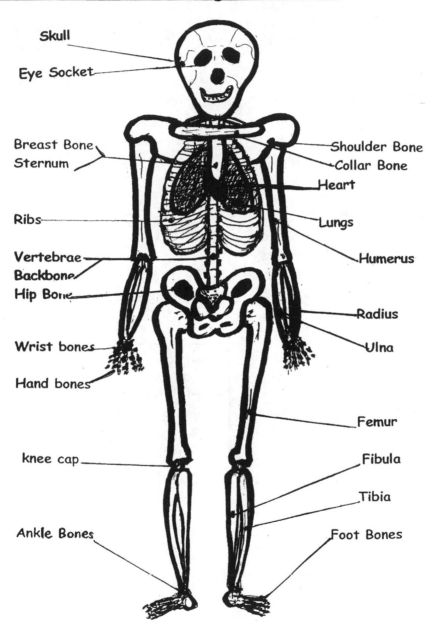

Skull

Eye Socket

Breast Bone
Sternum

Ribs

Vertebrae
Backbone
Hip Bone

Wrist bones

Hand bones

knee cap

Ankle Bones

Shoulder Bone
Collar Bone

Heart

Lungs

Humerus

Radius

Ulna

Femur

Fibula

Tibia

Foot Bones

Pattern for Making a Skeleton (2-6)

- Xerox a copy of the following skeleton pattern pieces on card stock paper.
- Cut out the bones.
- To put the pieces together, match the letters marked at the ends of each bone.
- Use a hole punch to make holes on the dots.
- Connect the bones using brass fasteners (found at any grocery, drugstore or office supply). When the skeleton is completed, the **bones** will move at the joints.
- Staple the 2 skulls together, using the markings at the top of the skull as a guide. Glue the **brain** in between the 2 skulls.
- On the bottom **ribcage**-glue the **lungs** and **heart**.

  (When adding the heart, lungs, and brain, talk about the function of each. A child's heart is approximately the **size of his fist**.

(See the guide to help in placement of bones)

Objectives-
1.   To learn related vocabulary
2.   To construct a child's sized skeleton
3.   To learn about vertebrates
4.   To identify some of main bones in the body
5.   To understand how bones protect soft organs

Vocabulary

✓ Skeleton	✓ Brain
✓ Vertebrate	✓ Joints
✓ Backbone	✓ Flexible
✓ Ribs –ribcage	✓ Protect
✓ Skull	✓ Bones
✓ Lungs	✓ Muscles
✓ Heart	

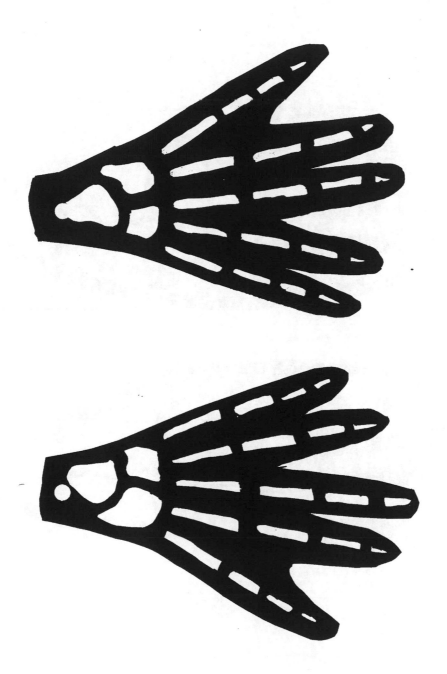

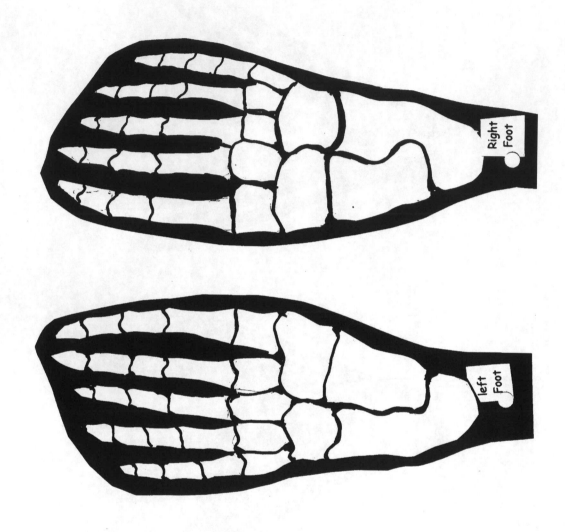

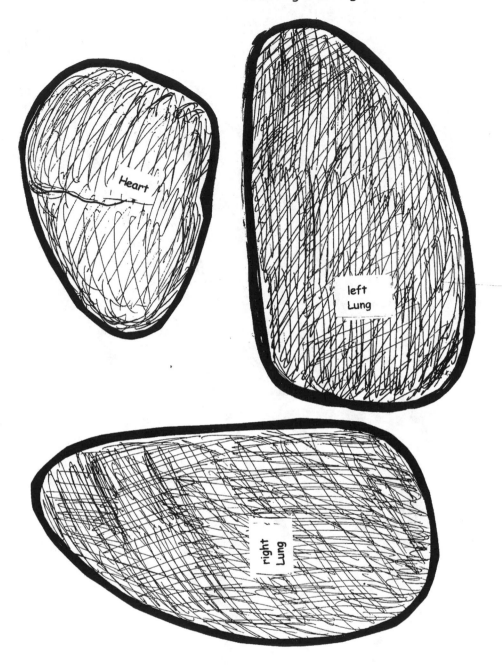

Mary Taylor Overton

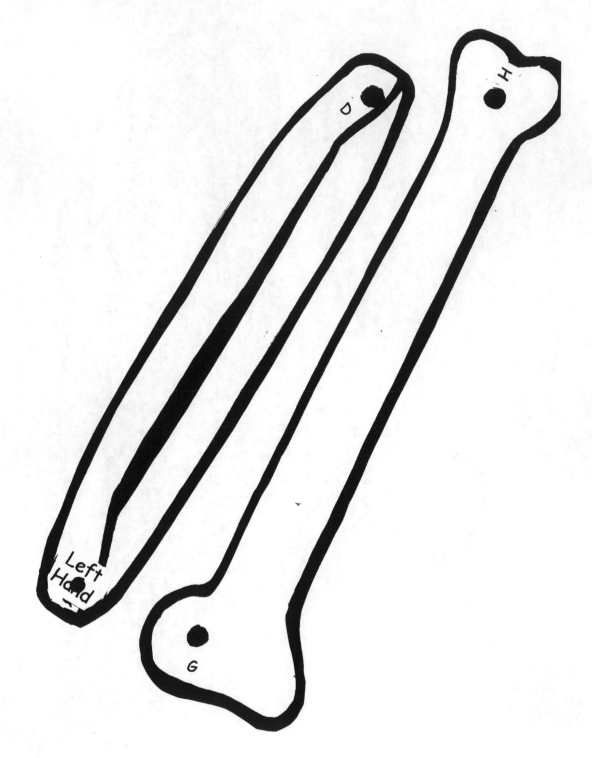

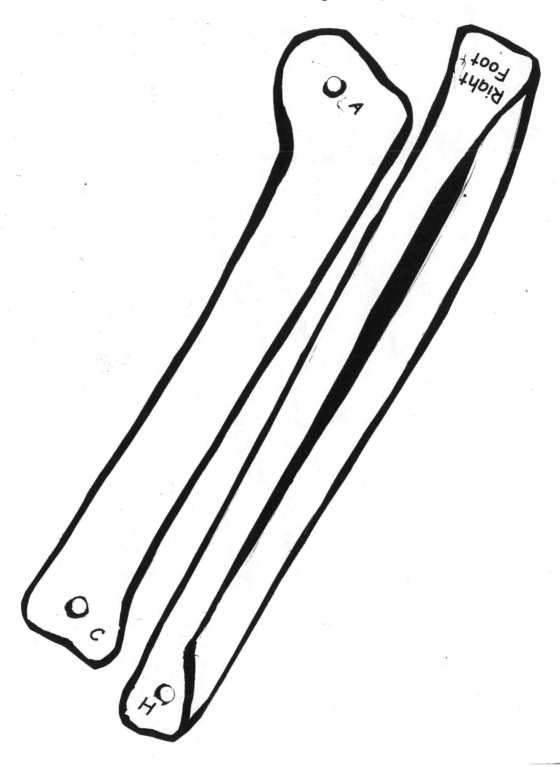

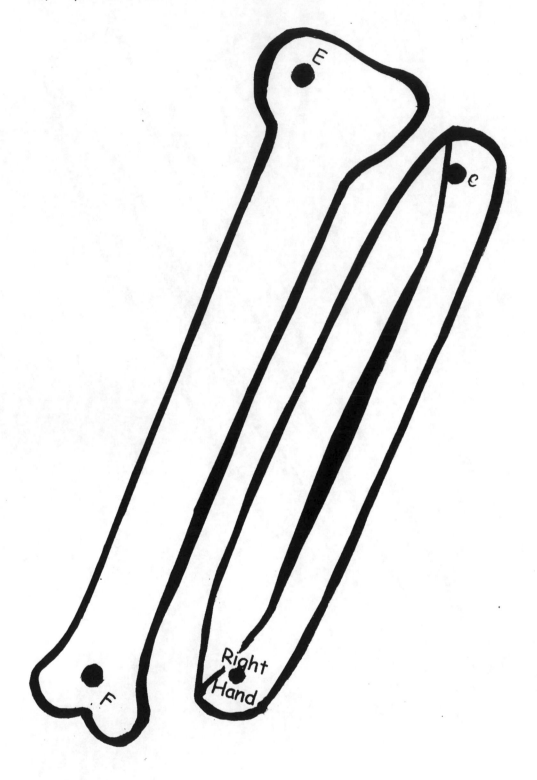

Glue vertebrae behind

V on the backbone for placement.

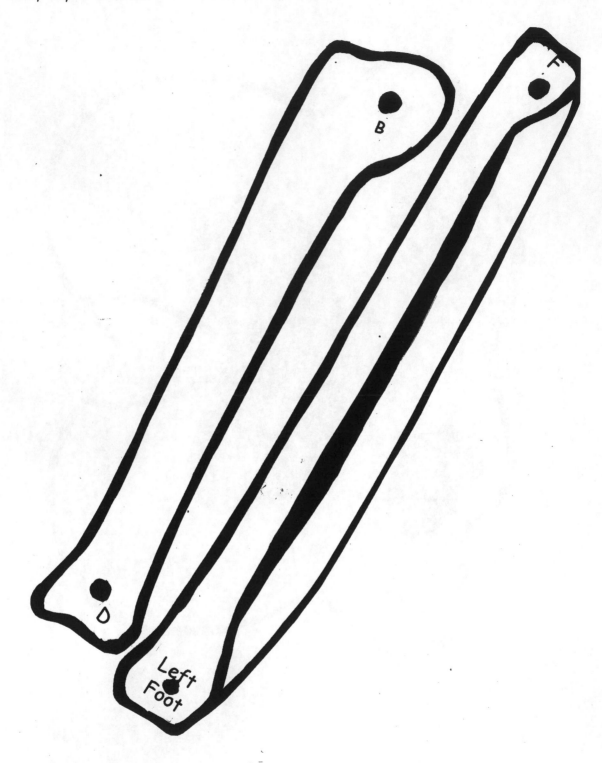

Vertebrae –

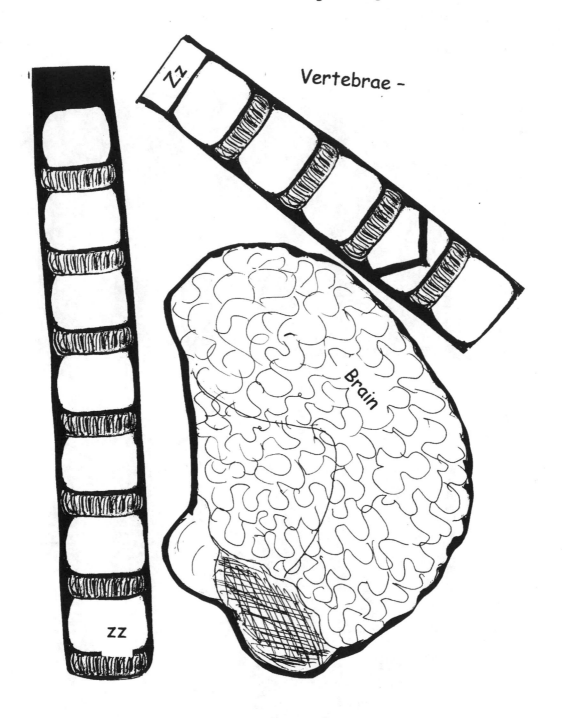

Brain

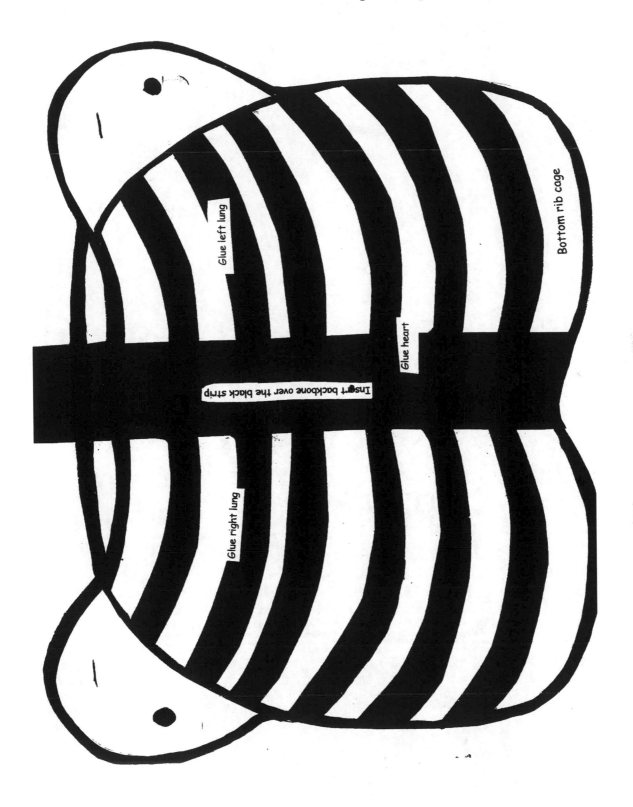

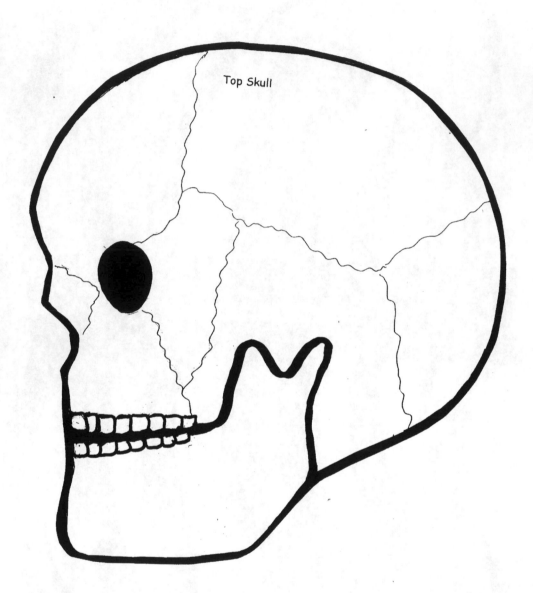

Top Skull

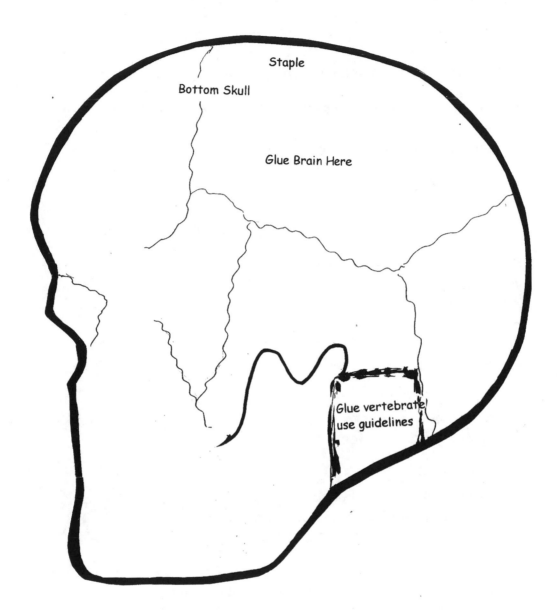

Mary Taylor Overton

Suggested Skeleton Activities-(3-6)

- 1. Read books about **skeletons**- animals and humans.

- 2. Use the skeleton to help your child to make discoveries about his/her own body.

- 3. Pose the skeleton and copy the pose.

- 4. Can you find the **longest** bone?

- 5. Can you find 2 bones that are the same?

- 6. Can you find a small bone?

- 7. In what part of skeleton do you see lots of small bones?

- 8. Use the skeleton to match bones in your body. "Can you find this bone on your body?"

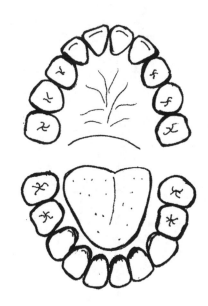

- 8. Can you find a bone that is **curved**? **Straight**?

- 9. Help your child to feel his/her **backbone** and the backbones of other family members. (Vertebrae)

- 10. If you have a dog or a cat, help your child to gently feel its **backbone**. Look at pictures of the bones of dogs and cats. Look at pictures of cat's and dog's skeletons.

- 11. Learn about other animals **with backbones-Vertebrates**.

- 12. Discover animals **without backbones-Invertebrates**

- 13. Read books about animals with **odd backbones- turtles**, **giraffes**, **snakes** and **swans**.

- 14. Discuss how the bones in the **skull** protect the soft delicate **brain**. Also discover how the **ribcage** protects the **heart** and the **lungs**.

- 15. Look at pictures of the body to discover how the **pelvic bone** protects the body's **soft organs**. Learn the names of some of these organs.

- 16. Do bending exercises to observe how **bones** move at the **joints**.

- 17. Cook a chicken's thigh with the drumstick attached. Carefully remove the meat from the bones and examine the **joint**. Discover how the joint connects to the bones and, and how the 2 bones move.

- 18. Exercise the joints in different parts of your body.

- 19. Collect the bones of chickens, pork, fish, or beef. Compare and contrast the bones. Compare the bones to yours. Look at pictures of animal skeletons and compare them to human skeletons.

- 20. Children are fascinated by the size of well-developed **muscles**. Use pictures to show how muscles move bones. What do you think would happen to your body if we did not have muscles? Invite a friend that works-out to showoff their muscles.

- 21. Remove the skin from the uncooked thigh and leg of a chicken. Observe muscles to see how they are attached to the bones.

- 22. **Mammals** have **lower jawbones** that can move. As you chew, feel the movement of your **jawbone**. Does your lower or upper jaw move? Use the skeleton to view the jaw bone.

- 23. Eat foods that are good for healthy bones. These foods that are high in **calcium**-include milk, ice cream, cheese, and yogurt. Discover other foods.

- 25. View x-rays of broken bones.

- 26. Discuss how bones get broken and how they mend.

Other Suggested Body Activities (3-6)

- 1. The **heart** is a **pump** that works just like any other pump. Use a fish tank pump to discover how a pump works. (4-6)

- 2. Human-**Blood types**- A, AB, B, O. Each blood type is also classified as negative or positive. Find out and remember your blood type. (5-6)

- 3. Everyone has his/her own set of **fingerprint** patterns, no two prints are alike. Use an ink pad to make fingerprints. Get closer look by using a magnifying glass. Put your prints in a special place. For security reasons, most police departments will fingerprint your child. (4-6)

- 4. Learn how your **lungs** work. (4-6) Look at pictures of the lungs. Take deep breaths. Discover how to care for the lungs.

- 5. Identify other body parts. (2-6)

- 6. Learn more about the heart. A child's heart is about the size of his/her fist. Find things around the house that are the same size. An adult hearts is about the size of an adult's fist. Compare the two. (3-6)

- 7. Use a stethoscope Listen to your heart beat, before and after exercise.

Teeth
Parent Background (2-6)

- Children are always interested in their teeth and can't wait to loose one. To the young child the loss of a tooth is a, "right of passage."
- A child is usually 4 or 5 years old before loosing a tooth.
- Teeth that fall out are called **deciduous teeth**, sometimes referred to as **baby teeth** or **milk teeth**.
- Deciduous teeth give way to **permanent teeth**. Permanent teeth are not replaceable.
- By the time that a child is 6 years old, he/she has 20 teeth.
- Adults have 32 teeth.
- Teeth are apart of the **digestive** system. Their main purpose is for tearing, grinding, and mashing food so that it can be safely swallowed.
- As food is chewed digestive juices in the mouth begin to break it down for further digestion.
- Teeth also help in the pronunciation of words.
- Many animals use their teeth to defend themselves and to capture food.
- However, humans do not **usually** use their teeth for defense or capturing food.
- Mammals have a variety of specialized teeth. The size, distribution, and number of teeth depend upon the diet.
- **Walrus, elephants**, and **wild boars** have tusks. **Tusks** are teeth that have grown longer than the animal's other teeth.
- Some animals such as **sharks** are able to replace lost teeth.
- Different species of fish have a variety of different kinds of teeth.
- **Turtles** do not have teeth. Their gums are sharp and hard.
- **Birds** do not have teeth. However, they are born with an **egg tooth** that is used for aiding the baby bird's exit for its eggshell.
- Goats are **herbivores** (vegetable eaters). They do not have teeth upper

front teeth. They have hard gums and nimble lips for pulling grass. Their lower strong back teeth are good for chewing fibrous foods. Goats have 4 chamber stomachs. Food is quickly swallowed into the 1st stomach for storage. It then enters the second stomach where it is softened into a curd. The curd is returned to the goat's mouth to be chewed, swallowed, and digested. This process is known as **chewing curd**. Animals that chew their **curd** are called **ruminants- deer, buffalo, mountain goats, cows,** and **sheep**.

- **Rodents, mice, beavers** and **rabbits** have teeth that are continuously growing. To keep their teeth from growing too long these animals must keep chewing hard materials.

Suggested Teeth Activities (2-6)

- 1. Read books about teeth and their care. (2-6)

- 2. Visit a dentist for a check-up. Ask your dentist to save an adult's extracted tooth for you. Compare it to one of your lost teeth. (2-6)
- 3. Count your teeth. (2-6)

- 4. Examine the teeth of an adult. (2-6)

- 5. Compare your teeth to those of a newborn or a younger child. (2-6)

- 6. If you have a family pet examine its teeth. (2-6)

- 7. Research how **goats, cows,** and **deer** use their teeth. (4-6)

- 8. Visit a farm or the zoo to see **goats, deer, cows,** or **buffalo**. Look carefully to see if they are chewing their curd.

- 9. Try saying words that require the use of the teeth-Example- teeth, nose, eighth, tight, kiss, kid, sunny, zoo, goat, love, think- Think of other words. (4-6)

- 10. **Beavers** are master builders. These animals have orange teeth, not white. They use their powerful teeth for cutting down trees as well as for dragging logs. Learn all about these little animals. (2-6)

- 11. **Elephants** have 2 long teeth called **tusk**. These tusks are not used for eating, but for digging. Elephants are either right or left tusked. From use, one tusk is usually shorter. Not all elephants have tusks. Research the African and the Asian elephants to find out which one has tusks. (2-6)

- 12. **Walrus** use their tusks to defense and digging in the mud in search of shellfish. Investigate these animals. (2-6)

- 13. Identify the **narwhal**. This whale has a long protruding tooth. (2-6)

- 14. Visit a pet store to observe rodents-**hamsters**, **mice**, **gerbils,** and **rabbits**. These animals have teeth that are continually growing. To control the growth of their teeth they must keep chewing or gnawing. Find out what kinds of foods are eaten by these animals. (2-6)

- 15. Very few animals have a more powerful jaw grip than the **alligator** and **crocodile**. These animals do not use their teeth for chewing. They tear the food or swallow it whole. Learn how these animals depend upon their teeth and jaws for hunting. (2-6)

- 16. Turtles do not have teeth; however, they eat a variety of different foods. Land turtles are usually **herbivores** (plant eaters) but, will include some meat in their diets. Water turtles are **carnivores** (meat eaters). Finds out how and what turtles eat. (3-6)

- 17. Brainstorm things that you must use your teeth to eat-Bite an apple, chew a piece of steak, corn-on-the-cob. Brainstorm things that can be eaten without using your teeth-ice cream, lollipops, drinks, mashed potatoes, and soup. (3-6)

- 18. Sharks replace broken or lost teeth. Learn about sharks. (2-6)

- 19. Learn how baby birds use their egg tooth. (2-6)

- 20. Birds do not have teeth. How do they eat? (2-6)

Top of the Food Chain

The Food Chain

Energy From the Sun

Mary Taylor Overton

The Food Chain –The Transfer of Energy
Parent Background (2-6)

- "Eat your spinach." "Fresh fruits make you grow." "Orange juice is full of vitamins." Your child will enjoy learning how energy gets into his/her body.
- All energy on earth begins with the sun.
- **Plants** and **animals** must have energy, without it they would not be able to carry out their daily activities.
- They depend upon the sun's light for **energy**.
- For their bodies to use this energy it must first get inside.
- This is accomplished through a complex partnership chain between the sun, the earth's green **plants**, and its **animals**. This process is known as the **transfer of energy** or the **earth's food chain**.
- All living things get their energy from the food that they consume.
- Animals are mobile and able to hunt for their food.
- Plants are stationary; therefore, they must produce their own food, or live off of other organisms.
- **Green plants** are the earth's link to the sun's light energy.
- These plants are the direct **receivers** and **dispensers** of the sun's energy.
- **Chlorophyll**, a green substance found in the leaves of green plants, absorbs the sun's light energy. This process is called **photosynthesis**.
- Powered by this light energy a chemical reaction takes place within the leaves.
- **Carbon dioxide** (a by-product from humans and animals that is released into the air, as they breathe), **minerals**, and **water** are combined within the leaf's structure to produce a sweet watery glucose sugary plant food.
- This energy-enriched food is called **nectar** in flowers and **sap** in trees.
- Plants use the food to grow, reproduce, and carry out other functions.
- Unused food is stored within the plant's **leaves**, **stems**, **roots**, **seeds**, **fruits**, and **flowers**, in the form of **sugars**, **starches**, and **proteins**.
- The earth's animals take in the energy stored in plants both **directly** and **indirectly** through the foods they eat.
- This is known as the **earth's food chains**, **the flow of energy**, or **the transfer of energy**.
- When humans or animals eat plants they are taking in energy **directly** from the plants.
- Animals take in energy **indirectly** when it eats another animal. An example: when we eat a piece of chicken we are taking in energy through the food that was eaten and store in the chicken's body.
- **Most of earth's food chains do not involve humans**. The majority of the food that is eaten by humans is farmed grown; this includes meat, fish, fruit, and vegetables. Food fed to farm fed animals is scientifically developed to ensure that the animals take in foods that are filled with vitamins and minerals.
- When animals and humans consume food their bodies immediately begin to prepare the food to be used as energy. The food is mixed with oxygen and

digestive juices, causing a chemical reaction to occur. The body burns the chemically produced **fuel** as heat **energy** that the body uses to carry out its daily activities.

- Animals living in the wild often fall prey to predators. Each time this happens the energy is transferred from one animal to the next. However, in the transfer the energy looses some of its strength. Therefore, to get enough nutrients these animals must consume larger amounts of food.

Animals are classified as:

✓ **Herbivores-** plant eaters
✓ **Insectivores-**insect eaters
✓ **Omnivores-** meat and plant eaters- including humans
✓ **Carnivores-** meat eaters
✓ **Parasites-** animals that live off of plants and other animals by using the energy produced by its host

- **Small herbivore animals** are at the bottom of the **food chain**. When these **herbivore** (plant eaters) animals eat the stored energy in the plant is transferred into their bodies. Small herbivores become a food source for an array of **small** and **large carnivores** (meat eating animals).

- Each time one animal falls prey; the energy within its body is transferred to its predator's body.
- **Larger carnivores** depend upon smaller animals as their energy source.
- An animal's size does not always determine what it eats. Large animals like **elephants, hippos**, and **apes** are **herbivores**-plant eaters.
- Small animals such as **spiders** and **ladybugs** are **carnivores**- meat eaters.
- Animals that are not usually hunted or eaten by other animals are at the **top of their food chains**. Owls, elephants, tigers, orca whales, polar bears, and humans are at the tops of their food chains.
- The **earth's food chain** is complex. The specific kind of foods eaten by an animal is directly related to the food source available at any given time. Land turtles are mainly herbivores (plant eaters) will occasionally include earthworms in their diets. These results in thousands of interconnected ever changing complex **food webs** between plants, animal herbivores, carnivores, and omnivores, living together in an ecosystem.
- Food webs can be found in all **ecosystems**-parks, oceans, woodlands, meadows, rain forests, gardens, deserts, lakes, ponds, streams, beaches, jungles, and the North and South Poles.
- The plants and animals living within any ecosystem are dependent upon each other for survival. Animals are dependent upon the survival of the plants, and plants are dependent upon the animals surviving.
- Some food chains are very short, while others are far more complicated.
- A **sample** of a typical food chain is:
➢ A **woodland food chain** might be the sun's energy–plant–ant–spider–frog–snake–owl.
➢ An **ocean food chain** begins -**sun-plankton** (small microscopic plants that float on the water and contain chlorophyll), makes its own food.

Zooplanktons are microscopic animals floating on the water that eats the plant **plankton**. Larger animals eat smaller ones. This continues until an animal at the **top of the food chain** consumes an animal.

➢   **Sample food chains –**
- Sun-plants- aphids-beetle bug- spider-frog- wolf
- Sun-plankton-zooplankton-small fish- larger fish-seals-polar bears
- Sun-plant-elephant
- Sun-plant-earthworm-robin-snake
- Sun-corn-pigs-man
- **Even in death**, all living things continue to be linked in the earth's closed ecosystem. **Decaying** plants, animals, and their eliminated wastes contain energy, minerals, and nutrients that must be returned to the soil to be recycled. This process is known as **decomposition**.
- Decomposition is the breaking down of complex animal and plant body tissue into chemical forms to be absorbed back into the soil. These materials will be reused by surrounding plants to make more food.
- Above and below the ground animals and plants known as **decomposers**, are feeding on the tissue of dead and decaying animals and plants.
- Without decomposers, dead things would never go away, but would remain on the ground forever.
- Decomposers set in motion series of chemical reactions that will result in the nutrients in a dead body being broken down and eventually absorbed back to the soil.
- **Earthworms, bacteria, mold, slugs, fleas, insects, larvae, flies, snails, mushrooms, millipedes, fungi, ants, slugs,** and **maggots** are just a few, of the many decomposers. Each ecosystem has its own decomposers.
- Once decomposition has been completed, the minerals and nutrients in the tissues of the animals and plants can be absorbed back into the soil, where they can be recycled by the surrounding plants. This is the same principal as the gardener when he mixes cow manure waste to enrich the soil.
- **Bacteria** make up the largest numbers of decomposers.
- ✓ They are in the air, water, ground, on the surfaces of everything, and even inside of the bodies of plants and animals.
- ✓ Bacteria exist in such large numbers that they are impossible to count.
- ✓ As soon as plants and animals die the bacteria inside and outside of the animal's body attacks its own tissues and begin to break them down.
- ✓ The first step in decomposition is the rotting and decaying the dead matter. The smell of rotting tissue will attract other decomposers to help in further decomposition.
- ✓ **Blowflies** lay their eggs on the rotting tissues. The eggs will hatch into maggot flies that will eat the rotting tissues of its host. The minerals and vitamins are later eliminated in the animal's waste and deposited on the ground. Again, decomposers attack the waste causing a chemical breakdown

to occur and allowing the minerals and nutrients to be absorbed into the surrounding soil.

- One of the busiest animal decomposer is the **earthworm**. All over the world millions of earthworms are busy eating dead leaves and debris and eliminating rich waste into the soil.

Vocabulary

-

Food chain
- Sun
- Energy
- Herbivores
- Carnivores
- Omnivores
- Prey
- Predators
- Top of the food Chain
- Bottom of the food chain

Decomposers

Suggested Food Chain Activities-

- 1. Read books about the sun's energy. (2-6)

- 2. Find out how plants produce their food. ($3\frac{1}{2}$-6)

- 3. **Make a Food Chain Game.** ($3\frac{1}{2}$-6)
  Use the food chain cards to tell food chain story. Make pictures on index cards. Example-On one card; draw the **sun**- on the 2nd card draw **corn**- on the 3rd card draw **a chicken**, and a on the 4th draw a picture of your child.

✓ Start with the **sun**.
✓ The **corn** plant grows using the sun's energy. Now the energy is inside of the corn.
✓ The **chicken** eats the corn and the energy is inside of the chicken.
✓ Your **child** eats the chicken and the energy is transferred to the child's body. The body chemically burns the energy as the child goes about his/her daily activities. Discover other food chains and make other cards to add to the game.

- 2. **Make other food chain cards**- (3-6) (Keep in mind the skill level of your child)

Examples- Use some of the sample food chains above.

- Sun – spinach- me
- Sun – apple- me
- Sun – earthworm- frog- snake- hawk
- Sun- grass- cow-milk-me
- 3. Make-up funny food chains.

- 4. At mealtime, help your child to explore the food chains of the different foods on his plate and in his glass? (sun- spinach- Me) (sun-grass-cow produces milk-Me) (Sun-plant-fish-Me) (3-6)

An Ocean's Food Chain Card Game (3-6)
- ✓ Plankton (a small green plant) floating on the surface of the water uses the sun's energy to make its own food. Plant plankton is eaten by floating animal plankton.
- ✓ A small fish eats the plankton and in turn an even bigger fish eats the smaller fish.
- ✓ The chain continues with the smaller fish being eaten by a larger or stronger fish.
- ✓ The chain stops when a fish or another animal at the top of the food chain, living in or near the water eats the fish.

- 5. Visit a pick your own vegetable or fruit farm. Here children can see fruits and vegetables growing and can pick some energy rich plants for dinner. (2-6)

- 6. Plant a small vegetable or herb garden. If space is a problem, plant a single tomato or green pepper plant. As the plant grows, talk about how the sun's energy is being used by the plant to grow. Also, discuss how the sun's energy is being stored inside of the plant. Enjoy the vegetables for dinner. (2-6)

- 7. Research your favorite animals, to determine what it eats and which animals are its **predators**. (4-6)
- 8. Discover a food chain in your backyard. (4-6)

- 9. Make a list of animals that are at the **top of their food chains. (Bears, owls, eagles, tigers, elephants, humans** etc.) (4-6)

- 10. Find out about animals that are at the **bottom of the food chain. (Most insects, Earthworms, crickets, mice)** (4-6)

- 11. Discover animals that eat only one kind of food, such as the **giant panda**, that eats only bamboo plants. The **blue whale** eats krill exclusively. (4-6)

- 12. Discuss what would happen if all the insects in the world were to die? Make a food chain that involves insects. Remove the insects. Now what happens to the food chain? (4-6)

- 13. As you read a good night story, decide if the animal in the story is an **omnivore, herbivore** or **carnivore.** (4-6)

- 14. Read about **herbivores, carnivores** and **omnivores.** (4-6)

- 15. Find out which animals are a part of the **Arctic** and **Antarctic** food chains. (4-6)

- 16. We usually think of **carnivores** (meat eaters) and as being only animals. However, there are hundreds of plants that are **carnivores-** meat eaters, these plants dine on insects. Read books about **Venus FlyTraps, pitcher plants, bladderworts**, and **sundews.** (4-6)

- 17. Discover a food chain in your neighborhood park. (4-6)

- 18. An animal's size does not always determine what it eats. **Elephants** are herbivores plant eaters. To provide its huge body with energy, everyday it must eat many pounds of plants. Research other large herbivores.–**hippos, apes, giraffes.**(4-6)

- 19. The blue whale and the whale shark, two of the largest animals on earth dine on small shrimp animals called **krill.** Learn about the feeding habits of these and other whales. (3-6)

- 20. Take a walk through the park; turn over a dead log to see what you can find. You should discover some **decomposers** at work. (4-6)

- 21. Set-up a fish aquarium. Add fish and snails. **Snails** are water decomposers. Observe how these little animals keep the bowl clean. Research other decomposers that live in the oceans or streams. (4-6)

- 22. Examine the leaves of plants and trees to find signs that an animal has been feeding on it. See if you can see the animal. (3½-6)

- 23. Setup an **earthworm** habitat to observe. $(3\frac{1}{2}-6)$
See invertebrates for more activities for earthworms.

Mammals

Mary Taylor Overton

MAMMALS
Parent Background(2-6)

• Mammals-The word **mammal** comes from the Latin- **Mamma**-means breast or nipples. Female mammals have **mammary glands** that produce rich nutritious milk to nourish their young.

• This includes such animals as **horses, pigs, whales, dogs, cats, mice, goats, rabbits, hamsters, lions, bats, elephants, bears**, and hundreds of other animals.

• Humans are classified as **mammals**. What sets man apart from other mammals is his **large brain** that enables him to think, and to engage in complex **problem-solving**.

✓ Humans have a **complex language** communication system.

✓ They also have excellent use of their opposable thumbs. Their thumbs allow them to accomplish complicated task using their hands.

✓ Man's skeleton bone proportion and his large toe placement on his gives balance and the ability to walk upright on 2 feet.

• All mammals have **backbones** or **vertebrae**.

• Mammals have internal skeletons.

• Each species has a body structure that is specifically designed for surviving in its environment.

➢ Horses, large cats, and wolves etc. have powerful limbs for running.

➢ Rabbits, deer, and elk have agile feet and legs to run and dart to safety.

➢ Whales and other sea mammals have flippers for speedy travel through the water.

• Mammals give birth to **live babies**. The exceptions are the **duckbill platypus** and two other species of mammals living in Australia and New Guinea that lay eggs. Except for this feature these animals exhibit the same characteristics as other mammals.

• **When mammals are born**, they are helpless and must depend upon their mother for survival. In some species, both the males and females work together to care for the young.

- The length of time young remain in their mother's care varies from species to species. In some species females remain together for life.
- Parent(s) also transmit skills from one generation to another. They teach their young the same skills taught to them by their parent(s). This includes getting along within a species, hunting for food, and remaining out of harms-way.
- Many species live in a group and have a **social order**. Elephants and monkeys have a definite social order.
- Mammals have some means of communicating within their species.
- Communication consists of **talking, growling, grunting, howling, squeaking, wagging** their **tails, hugging, patting, stroking, grooming**, or **facial expressions** are a few.
- Mammals eat a variety of different foods. Some are **carnivores** (meat eaters); this group includes **dogs, cats, bears, foxes, weasels, otters, seals, walruses**, and some **whales**. Others are **herbivores** (plant eaters). They have digestive systems that are designed to breakdown tough fibrous materials. **Herbivore** mammals include **rabbits, squirrels, deer, elk, sheep, horses, cows, pigs, elephants, giraffes, mice, beavers**, and **panda bears** etc. Many of these animals will occasionally include some meat in their diet.
- Humans are most often grouped as **omnivores** (meat and plant eaters).
- Mammals are **warm-blooded**. They are able to produce body heat to maintain a constant inner body temperature.
- All mammals have **hair or fur**. Some mammals have lots of hair or fur. Whales, porpoise, and elephants have only small amounts of hair or fur.
- Various patterns the on each specie's skin gives camouflage, protection, and help in identification.
- Examples:
- **Zebras** live in together in a herd producing a sea of stripes; this makes it difficult for predators to see individual zebras.
- **White-tailed deer** have brown coats that are excellent for camouflaging behind the brown coloring of trees.
- **Lions** use their tan brown coats for hiding in the tall brown grasses.
- **Polar bears** have white fur that blends in with the surroundings.
- Mammals have **lungs and breathe air.**
- Most large mammals do not build shelters, but seek it wherever they can. However, some pregnant mothers prepare a birthing place. A bear will often dig out a den in the side of a hill.
- Mammals live above and below ground, in trees, under trees, in oceans, rivers, and hidden in grass.
- Some species of small mammals build or dig complicated above and underground **burrowing** systems.
- **Beavers** and **muskrats** build strong structures above and below water
- **Mammals** have a variety of different kinds of limbs.-(hands, hooves, padded feet, claws, flippers, toes, webbed feet)

- **Bats** are the only mammals that can fly, but their bodies are covered with fur just like other mammals.
- **Hibernation**-Contrary to belief, mammals do not **hibernate** because they cannot withstand the cold weather. During the winter months food is hard to find to conserve energy many mammals slow their daily activities.
- Mammals like bears, squirrels, skunks. and chipmunks spend time in their dens sleeping, and resting in a **semi-hibernation** state. These animals will awaken to eat and eliminate their waste.
- Groundhogs are **true hibernators.** Their body temperature lowers. Their heart rate and breathing slows and they spend the entire winter sleeping.

OBJECTIVES
1. To learn related vocabulary
2. To learn the characteristics of mammals
3. To compare and contrast mammals with other animals

VOCABULARY
- Hair
- Fur
- Habitats
- Rich milk
- Nurse babies
- Born alive
- Breathe air-lungs
- Females- males
- Mammals
- Vertebra
- Backbones
- Limbs
- Camouflage
- Carnivore
- Herbivores
- Omnivores
- Predators
- Prey

Suggested Mammals Activities (2-6)

- 1. **Mammals**-cats, dogs. and humans can be observed daily.

- 2. **Gerbils, hamsters** and **white mice** are available from pet stores. These animals make excellent pets and can offer your child an opportunity to see these mammals interact. Buy 2 of the same sex, unless you are interested in supplying animals for the neighborhood. (2-6)

- 3. In the park or your backyard observe squirrels at play. You can attract squirrels by feeding them. Place food in a safe place, away from dogs, cats, and children. Sit quietly and wait for a squirrel. What are the squirrels doing? Do they seem to be playing together? Discuss why it is dangerous to catch wild animals. (2-6)

- 4. Visit a pet store, zoo, or a farm to observe **mammals**. (2-6)

- 5. Observe 2 dogs at play; do they seem to be communicating with each other? If so, how? (2-6)

- 6. Read books that show how mammal babies are cared for by their parent(s).

- 7. Visit a zoo to find mammals with babies. Observe how each mother cares for her young? If there are other animals in the cage, how do they react to each other? To the baby? (2-6)

- 8. Read stories about mammals. (2-6)

- 9, Compare **mammals** to **birds**. How are they alike? Different? (3-6)

- 10. Select a mammal to learn all you can about it. (3-6)

- 11. Draw pictures or make clay models or puppets of mammals. (2-6)

- 12. Discover how **whales** breathe. They have a **blowhole** in the top of their heads. Compare whales to other mammals. (2-6)

- 13. Find mammals that are **carnivores** (meat eaters), **herbivores** (plant eaters), or **omnivores** (plant and meat eaters). (4-6)

- 14. Draw a mammal and **camouflage** it in its surroundings. (3-6)

- 15. Pretend to be a mammal, camouflage yourself so that you won't be seen. (3-6)

- 16. Look through the toy chest to identify the mammals. (2-6)

- 17. Go on a "**Mammal Hunt**," ride. Keep track of the number of mammals identified. Accept pictures of mammals on billboards or signs. (2-6)

- 18.   Children like knowing about the **strangest**, the **largest**, the **smallest**, the **fastest**, and the **strongest** mammal. Read books to discover fascinating things about some of these mammals. (3-6)

- 19.   Breast-feeding is a **natural part of life**. If young children see babies being nursed they will understand the process. Nursing is again back in style, seek out a friend who is nursing. Visit a pet store to observe a mother mouse feeding her young. (2-6)

- 20. Learn about odd mammals such as **kangaroos**, **duckbill platypus**, and **bats**. (2-6)

- 21.   During the winter, some mammals go into a state of **hibernation** or **semi-hibernation**. Read more about hibernating mammals such as **bats, groundhogs**, **bears**, **chipmunks**, and **squirrels**. Find out how mammals prepare for winter. (3½-6)

- 22.   Some mammals remain in family units throughout their lives. Find out which mammals work together as a social unit- monkeys and lions. (2-6)

- 23.   Investigate man's closest relatives, the **monkey family**- apes, gorillas, orangutans, and chimpanzees. (2-6)

- 24. Learn the male, female and baby names of some mammals. (3-6)
- Examples:

Name	Male	Female	Young
Seal	bull	cow	pup
Panda	boar	sow	cub
Monkey	male	female	infant
Lion	lion	lioness	cub
Human	man	woman	baby
Horse	stallion	mare	foal
Gerbil	buck	doe	pup
Kangaroo	jack	jill	joey

- 25. Learn the names given to different groups of mammals: (3-6)
- A group of elephants is called a **herd**.
- Lions - **pride**.

- Porpoises-**pod**
- Leopards-**leap**
- Dogs-**pack**
- Kangaroos-**troop**
- Bears-**sleuth**
- Horses- **stable**
- Gorillas-**band**
- Foxes-**leap**
- Lions-**flock**
- Beavers-**colony**
  Discover other names.

- 26. Dramatize some of the many different ways that mammal's move- run, walk, or jump etc. (2-6)
- Jump like a **kangaroo**.
- Walk like an **elephant.**
- Hop like a **rabbit**.
- Stretch like a **giraffe**.
- Leap like a **deer**.
- Drag yourself like a **seal**.
- Swim like a **whale**; breathe through your blowhole.
- Hang like a **bat**.

- 27. Discover **nocturnal** (animals that are active at night) and **diurnal** (day) mammals. (3-6)

- 28. Find out about mammals that live underground-**chipmunks, badgers, groundhogs**, and **moles**. Discover what their underground homes are like. (2-6)

- 29. What species of mammals are indigenous to your state, country or continent? (3-6)

- 30. Read about **marsupials, kangaroos,** and **opossums**. (2-6)

- 31. Discover **Arctic mammals**-polar bears, **musk oxen, seals,** and **Arctic foxes** etc. (3-6)

- 32. Research mammals that **migrate**. (4-6)

- 33. Discover some of the many mammals that are **endangered**. This includes **elephants, lions, hippos, tigers, pandas, blue whales, great apes, sea cows,** and other **sea mammals**. (3-6)

- 34. Compare the largest land mammal, the **African elephant** to the largest sea mammal, the **blue whale**. (3-6)

- 35. Find out about mammals that have quills instead of soft hair or fur—**porcupines** and **hedgehogs**. (2-6)

Fish

Mary Taylor Overton

Fish
Parent Background
- Fish have lived on earth for millions of years.
- They can be found in ponds, streams, rivers, lakes, bays, and oceans.
- Fish are divided into 2 categories, bony and cartilage fish. Bony fish have skeletons made of bone and cartilage fish (sharks, rays and skates) skeletons are made of cartilage.

- ## Gills
  - › Fish breathe through **gills** located under the **gill covers** on each side of the head. The gills act as lungs by removing oxygen from the water. As a fish swims water enters through its mouth and is forced across the gills.
  - › The gills are covered with blood vessels that become bright red as they absorb oxygen from the water.
  - › The oxygen rich blood is carried to the heart where it is pumped through the fish's body. As the blood flows through the organs in the body, the oxygen is absorbed.
  - › Each cell in every organ uses the oxygen and releases carbon dioxide (a gas by-product produced by each of the organ's cells).
  - › The carbon dioxide is absorbed back into the blood where it is taken to the gills, and forced into the surrounding water.
- Fish are **cold-blooded**. The temperature of the surroundings water helps to regulate their body temperature. Fish are able to generate some body heat, but not as much as mammals and birds.
- They are a vital link in the earth's **food chain**.
  - › All **water food chains** begin with plant **plankton**, called **phytoplankton**. They are microscopic green plants that float on the surface of water. These plants are the primary producers of food for all water food chains. These small green plants contain **chlorophyll** that **absorbs** the sun's **energy** and uses the energy to produce food.
  - › Also floating in the water are microscopic animal planktons, called **zooplankton** that feed on plant plankton. Animal plankton consists of a wide variety of small animals, such as baby fish, and larva from many sea creatures. Plankton is a **rich** food source and is often called, "**nature's soup.**"
  - › Various water animals living in the ecosystem consume plankton.
  - › The diets of the many different kinds of fish living in any ecosystem varies.
  - › All animals living in a single ecosystem are connected to that system's food chains and food webs.
  - › Some fish are **herbivores**-feed upon small plants and plant plankton. Most fish are **carnivores**-meat eaters that feast on fish and sea creatures. There are many species that are **omnivores** feeding on both plants and other fish.
  - › Some species are **parasites**. These fish attach themselves to other fish and feed off of its host.

➤ An animal's size does not always determine what it consumes. Whale shark, the largest fish in the sea eats plankton.

- **Fins**-Fish move by the means of fins that are located on various parts of the body.
- The **tail fin** pushes the fish forward.
- The **pectoral fins** help in turning.
- The **dorsal fin** keeps the fish upright.
- Most fish have **swim bladders** located in their inner body cavity that act a flotation device, by helping to maintain body balance. It also keeps the fish from sinking. A swim bladder is like a balloon that the fish can inflate and deflate with air gases. To raise the fish inflates the swim bladder with air, this makes the body lighter.

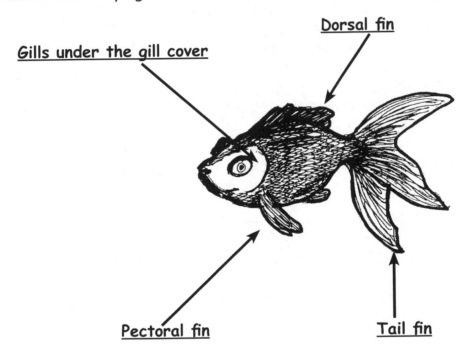

Gills under the gill cover

Dorsal fin

Pectoral fin

Tail fin

- **Scales**-Most fish are covered with **translucent scales** that protect their thin skin. The skin's color can be seen through the translucent scales.
- The skin color serves as **camouflage** from **prey** and **predators**, by helping the fish to blend in with its surroundings.

Vertebrates
- Fish have **backbones** and are classified as **vertebrates**.
- Backbone and ribs gives the fish its shape.
- Most fish lay eggs, a few give birth to live young.
- Baby fish are called **larvae** or **fry**.
- Some fish care for their young, most do not.
- Small species often gather together in what is known as, "**schools of fish**." There is protection in numbers.

- Fish have varying degrees of hearing, depending upon the species.

Senses
- There is a sensitive **lateral line** that runs down either side of the body. This line can detect sounds through water, even the slightest **vibration** can be felt.
- **Nostrils** are used for smelling not breathing.
- **Taste**-Fish taste in a variety of different ways.
- Fish cannot close their eyes. A clear shield covers and protects the eyes.
- They do not sleep, but engage in periods of resting.
-

Sharks are Fish Too!
- This family includes **sharks**, **rays**, and **skates.**
- There are hundreds of different species of sharks.
- Sharks have bones that are made of **cartilage** not bone.
- They do not have swim bladders. They have large livers that are full of oil. Oil is lighter than water, helping to keep the shark's body afloat.
- Most sharks are **carnivores** (meat eaters).
- The **whale shark** is the largest. It can grow to about 60 feet long.
- **Shark's** have bodies that are well suited for the **predators** that they are.
- They are **vertebrates**. Their skeletons are made of **cartilage** not bone.
- They do not have swim bladders and must keep swimming to keep from sinking.
- Sharks do not have traditional **scales**. Their bodies are covered with a rough texture much like that of course sandpaper.
- Most sharks give birth to live young. They are on their own as soon as they are born.
- Not all sharks are harmful to people. There are only a few species that will actually attack humans.

Suggested Fish Activities- (2-6)
Provide a wide variety of different kinds of picture books about fish.

The following activities have been divided into two parts. The activities are designed to give your child hands-on experiences.
- Part 1- Making discoveries using a fish purchased from the fish market
- Part 2- Observation of a live fish in its habitat.

Part 1
(Making Discoveries)
- We are always telling children that fish are covered with **scales**. The activities in part 1 are designed to give your child tactical and visual

experiences with the body make-up of a fish.

- Purchase a whole fish from the **fish market**. The scales, fins, head, and inside organs **must not be removed**. Ask the fishmonger for a female fish, there might be fish **eggs (roe)** inside. Eggs are small, round and covered with a jelly-like substance. Some fish have only a few eggs while others have thousands. (2-6)

Observation: (2-6)

- Provide a good powerful magnifying glass.
- 1. What is the shape of the body? (Describe) Take time to carefully examine your fish. Look at books of fish and compare your fish to the shapes of other fish. How are they alike? Different? (2-6)

- 2. Trace around the body of the fish, using your fingers. (2-6)

- 3. Feel the scales and skin-"Why do you think fish have scales?" (2-6)

- 4. Pull off some of the scales. Use a magnifying glass to get a closer look. (2-6)

- 5. Note the color of skin with the scales/without. (2-6)

- 6. From the hardware store, purchase a piece of very course sandpaper. Sharks have skin that is much like this texture. Compare the smooth skin to the texture of the sandpaper. (2-6)

- 7. When touching fins, use caution-**fins have sharp ends**. (2-6)
- ✓ Spread the **pectoral fin**.
- ✓ Feel the tail fin.
- ✓ Move the tail fin as if swimming in water.
- ✓ Move the entire body to feel its flexibility.
  Compare the fins of several different species of fish.

- 8. Observe the **mouth**. "What do you see?" "How does it feel?" Use a mirror to compare your mouth to your fish's mouth. (2-6)

- 9. Compare the tongue to your tongue. (2-6)

- 10. Look for the teeth. Compare them to your teeth. Are they like yours? (2-6)

- 11. Can you find the nostrils? Nostrils are not used for breathing, smell only. How do you breathe? (2-6)

- 12. Compare your eye to the fish's eye. Touch your eye, feel its eye. How are they alike? Different? Look for eyelids. (2-6)

- 13. Read about how **gills** work. Discuss how the gills are used to remove oxygen from the water. Carefully lift the gill covers to see the **gills**. (4-6)
✓ "How do they look?"
✓ "What color are they?"
✓ "How do they feel?"
✓ "Show me how the gill covers move?"

- 14. Find the lateral line that runs down the sides of the body. Discover how these lines help the fish to hear. (3-6)

- 15. **Cook a whole fish** and carefully remove the flesh to reveal the bones. Try to keep the skeleton (**vertebrae, ribs, head**, and **tail)** in one whole piece. Examine the bones. How do they feel? Compare the bones to those of a chicken, beef, pork and to your bones. (2-6)

- 16. Older children (5-6) can **dissect** a fish to **discover** and **identify** the inner parts. Before dissecting, trace the outline of the fish onto paper. Your child can draw in the body parts as he/she makes discoveries.

- 17. Examine the fish to see how the **tail** and **head** are attached to the **backbone.** (2-6)

Part 2
- **Observing a Fish in its Natural Habitat** (2-6)
- Set-up a fish bowl or tank. Feeder gold fish cost about 25 cents each, at any pet store. For the purpose of comparison, get at least 2 different species.
- After setting up the tank, give your child time to **observe** the fish. You'll be surprised just how much will be learned simply, by watching the fish go about its daily activities. Always be available to answering questions. However, let your child form his/her own answers by allowing him/her time to think. An example: When your child asks, "How do fish swim?" Answer, "How do you think he swims?" "Let's watch".
Later, intervene by guiding him/her to make other discoveries and connections.
1. "What do you see?"
2. "Why do you think?"
3. "What is different?"
4. "How do you know?"

Observation (2-6)

- 1. Guppies are excellent for observing the **life cycle of fish**. These little guys give birth to live babies. Ask your fish dealer about their care.

- 2. "When you swim, how do you move your arms and feet?" "How does the fish swim?" Pretend to swim. (2-6)

- 3. "How does it use its **tail, dorsal,** and **pectoral fins**?"

- 4. "Are the fins ever still?"

- 5. "Does the fish swim backwards, sideways, or upside-down?"

- 6. "How does the fish turn?"

- 7. If you have more than one fish, compare the 2 by -size, marking, coloring, markings, gills, eyes. "How are they alike?" Different?

- 8. "Do they play?" "Do they ever bump into each other?"

- 9. "Show me how the **gill covers** move?" "Show me how you breathe?"

- 10. "What do you think would happen to the fish if no water was in the bowl?"

- 11. Observe the fish's mouth. "What do you see?" "Can you see his teeth?"

- 12. "What happens when you add food to the tank?" "How do fish know that there is food in the water?

- 13. "Can fish hear?" " How can we find out?" Tap on different sides of the tank, to see what happens.

- 14. "What happens when the fish comes to the **surface** of the tank?"

More Fish Activities (2-6)
- 1. Compare a fish to dogs, reptiles, mammals, insects, and other animals. "How are they alike? Different?" "How is a fish like you?"

- 2. Adding an air rock to the bowl will add oxygen to the water. "Do you think this is a good idea? Why do you think it is a good idea to add an air rock?" Use a straw to blow air into a clear glass containing water. What do you see? Why?

- 3. "Why do you think the water in the fish tank gets dirty?" This is an excellent time to talk about water **pollution** and the problems it causes for fish and other animals living in water. Read books about **water pollution,** note oil and chemical spills. (4-6)

- 4. Pretend to be a fish swimming in the ocean. Hide from your predators. What kind of a fish are you? Where are you hiding? Are you a **carnivore, herbivore** or an **omnivore**? What kind of food are you looking for?

- 5. (2-6) Make a copy of the **fish puzzle.** (See pattern at the end of the fish chapter). This puzzle is a great hands-on learning tool. Glue the puzzle on cardboard and cut it out. The cardboard will make the puzzle durable. (5-6) Choose a favorite fish and make your own fish puzzles.

- 6. Learn about the **Coelacanth.** This prehistoric fish was thought to have died out 70 million years ago. It was rediscovered off the coast of Madagascar. Locate Madagascar on the globe. Find out about this fish. (4-6)

- 7. **Sharks** are fish. View pictures of sharks. "What do you notice about their fins?" Compare their fins to other species of fish. Read books about sharks. Learn to identify **rays** and **skates**; they are members of the shark family. (2-6)

- 8. The **whale shark** is the largest **fish** in the ocean. Learn more about this fish. Measure 60 feet to show its length. Compare 60 feet to other things. (3-6)

- 9. The whale shark is an endangered fish. Discover other fish and sea animals that are endangered, threatened, or extinct. (3-6)

- 10. Use small fish crackers or gummy fish to create a **school of fish.** Eat one fish at a time. Find out the names of some species of small fish that are members of schools of fish. –sardines, herring(3-6)

- 11. Draw an outline of a fish, make several copies. Your child will enjoy creating and designing his/her own exotic fish. (3-6)

- 12. Visit a pet store to see some of the many different varieties of fish available. (2-6)

- 13. **Salt-water** fish tanks will support beautiful exotic fish. These tanks require lots of care. Visit the pet store to see one of these tanks and its exotic species of fish. . Discover what it means to be a **fresh** or a **saltwater** fish. (3-6)

- 14. There are thousands of food chains in the earth's, oceans, bays, streams, ponds, brooks, rivers, coves, and other waterways. Learn about

some of these **food chains**. (3-6)

- 15. Identify fish that live in the deepest part of the ocean. It is so deep that very little sunlight filters through. Learn about this section of the ocean. (4-6)

- 16. **Eels** look like snakes, but are fish. Discover how they differ from snakes. Many Asian fish markets sell eels. Take a visit to the market to see an eel. Purchase an eel and a fish. Compare the 2 fish. (3-6)

- 17. Discover animals that are called fish, but are not-**Starfish**, **jellyfish**, and **shellfish**. (3-6)

- 18. Learn about the fastest fish, the **Marlin**. (3-6)
- 19. Identify some **poisonous** fish. (3-6)

- 20. **Sea horses** are fish. Find out more about these fish and their cousins the **sea weed dragons**.

- 21. There are many different kinds of animals living in the sea. (**Crabs, lobsters, corals, octopus, jelly fish, sea anemones**). Learn more about some of these sea creatures.

- 22. Learn about strange looking fish like **lionfish**, **clown fish**, **puffer fish**, and **ocean sunfish**. (3-6)

- 23. Identify **sea mammals**-whales, sea lions, seals etc. Compare them to fish. (3-6)

- 24. **Flounder** is one of many species of **flatfish**. These fish are flat and have both of their eyes on the same side. When flounders begin life, they are born looking just like other fish. They swim in the same position as other fish. One eye is on each side of the head. However, as they grow, the flounder gradually begin to turn to one side. One eye slowly moves to either right or the left until both eyes are on the same side of the head. The fish sink into the bottom of the water, where it hides in the sand, camouflaged from predators. From its bottom position it uses its eyes to look up for food and predators. Visit a fish market and purchase a flounder fish to examine. Compare it to another fish. Note the position of the mouth.

- 25. Find out about species of fish that can be found in your state's local waterways. (4-6)

- 26. **Flying fish** appear to be flying as they leap out of the water trying to escape from their **predators**. These fish have large pectoral fins that it spreads, as it jumps, this causes the wind to catch the fins. This results in the fish appearing to fly through the air. They can glide over 150 feet. Measure 150 feet. Learn more about these fish. (3-6)

- 27. Sort and graph gummy fish by colors. (3-6)

- 28. Read story books about fish- "**Rainbow Fish**," "**Swimmy**," " **Big Al**". (2-6)

Fish Puzzle-
Copy the puzzle and cut out-Use the dotted lines for the placement of fins.

Birds

Birds-

Birds are always available for observing. They are found all year around, in your yard, parks, in the city, on lakes, ponds, or country sides. Birds are fascinating to children, because they can do amazing things like lay eggs and most can fly.

Parent Background

- Birds have been on this earth for million of years. Many scientists believe that birds are the closest living relatives of the dinosaur.
- Birds are found on all **7 continents**.
- There are hundreds of different species.
- They come in every color in the rainbow.
- Most birds can fly.
- Birds are the most mobile animals on earth.
- They have the ability to travel all over the world.
- Some birds like **penguins** and **ostriches** cannot fly.
- **Children enjoy learning about unusual birds:**
✓ Ostriches, biggest
✓ Hummingbird, smallest
✓ Penguins, flightless
✓ Eagle, strong and powerful
✓ Owls, good hearing
i

Characteristics of Birds

- Birds eat a variety of foods.
✓ **Carnivores**- Meat eaters
✓ **Omnivores**- Meat and vegetable eaters
✓ **Herbivores**-Vegetable eaters
✓ **Insectivores**-Insect eaters
✓ **Birds do not have teeth**- They have a special organ called a **gizzard** for grinding food.
- Birds are **warm-blooded** and are able to maintain a constant internal body temperature. The internal temperature of birds is very high, around 100 degrees.
- When food is consumed, it is chemically converted into fuel and is used by the body as heat energy. Birds must consume large amounts of food to keep their body heat temperatures at its required high level. There is no truth to the statement, "eats like a bird." Birds are big eaters; some eat their weight in food everyday.

- **Bills**-All birds have **bills** or beaks. The shape of the bill depends upon the kind of food eaten, as well as the bird's lifestyle. Birds use their bills as hands. Bills are used to hold, to carry things, to build their nest, for defense, to feed their young, to groom, and to capture food.

- **Birds of prey** like **eagles** and **hawks** are meat eaters (carnivores) have a hooked bill for tearing. The **scarlet ibis** has a curved, sharp bill used for spearing frogs, fish, and other small animals. **Pelicans** have a bill for scooping fish from the water. **Robin's bills** are shaped for pulling and holding earthworms. The **flamingo's bill** is used for digging in mud and as a strainer for separating shellfish from the water and mud. The **woodpecker's** bill is strong for digging into hard wooden tree bark. **Hummingbird's** bills are long, thin, and hollow for sucking nectar from flowers.

- **Feathers**-All birds have feathers, called **plumage**.
- Each species has its own distinctive pattern and color. Patterns and colors help members of the same species in identifying one another.
- In many species males are more colorful than females. Males often use their colorful plumage for attracting a mate. Females of a species often use their dull feathers as **camouflage**, as they nest.
- Dull brown feathers camouflage owls hiding in the top of trees. Birds living in tropical forests have colorful feathers that can be used to hide among the flowers.
- Feathers take a beating and must be replaced. Each year birds loose and replace some of their feathers. This process is called **molting**.

- Feathers require constant care. Birds spend hours pulling rich oils from their skin on to their feathers. This is called **preening**.
- Feathers provide insulation against the hot, cold, and wet weather.
- Soft down feathers overlap, trapping heat close to the skin. During hot weather, these same down feathers are fluffed to allow cool air to circulate to the skin.
- In some species, baby birds are born without feathers. Hatchlings are usually born covered with soft down feathers. These young birds cannot fly until the down feathers molt (loose feathers) and are replaced by strong **flight feathers**.

- **Feet**-Birds have 2 **feet**. The shape and size of a bird's foot also depends upon the bird's lifestyle and feeding habits. Webbed feet found on water birds, are designed to move large amounts of water. **Eagles** and other **birds of prey**, have sharp **talons** on the ends of their toes, for grasping prey. Ostriches cannot fly, but have huge feet and strong legs, that are used for running and defense. **Flamingos** have feet that are suited for spending much of their time in shallow muddy water looking for food.
- Most birds have **4 toes** on each foot. The placement of the toes depends on their use. **Perching birds** like **robins** use their toes for griping and balancing on branches. The **woodpecker's** toes are used for balancing as he hangs vertically on tree trunks.
- Birds have hollow bones, making their bodies lightweight and suitable for flight.

- Most birds build a **nest**, others use abundant nests, hollow trees, the ground, and some even lay their eggs in the nest with the eggs of other species of birds.
- In some species, male and females birds remain together for life.
- All birds hatch from **eggs**.
- Most baby birds are helpless and must depend upon their parent(s) for survival. Some birds like baby chickens are able to care for themselves soon after emerging from the egg.
- In most species, both males and females care for the young.
- Baby birds are usually born in the spring of the year. If it is a **migrating** species the young birds must be ready to travel by the fall.
- Just before the winter months some species **migrate** south. Here it will be easier to find food. Many migrate to their birthplaces to mate. Some birds like the **Arctic turn** travel thousands of miles each year.
- Scientists are not sure how birds navigate their migratory routes.

Suggested Activities for Observing Birds (2-6)

- 1. Read books about birds, include a field guide for identifying.

- 2. Think of famous birds. (Ex. Big Bird, Tweety bird, Thanksgiving turkey...) (2-6)

- 3. Become a **bird watcher** and observe birds in your yard. (2-6)

- 4. Keep a record of the birds that you find and identify. (4-6)

- 5. Get an inexpensive pair of **binoculars**. (3-6)

- 6. In the fall of the year watch for **migrating** birds flying in a **V formation**. (2-6) "Why do you think they fly in a V formation?"

- 7. Purchase a tape of **birdcalls**. Try to imitate the calls. Find pictures of the birds to match the calls. (4-6)

- 8. Take a bird, "**listening walk**". Listen to the sounds of birds singing. (2-6)

- 9. Go on a, "**nest hunt**". During the winter when the trees are without leaves, nest can be easily found. Do not touch the nest. Discover the many different places where they can be found. Birds living in the city have adapted their lives to survive there. They have discovered many unique places to build nests. See if you can find some of these places. (2-6)

- 10. A bird makes an excellent pet. (2-6)

- 11. Observe how birds **take off** and **land**. (2-6) Compare landings. (5-6)

- 12. Put a **birdbath** in your yard. (2-6)
- 13. Purchase feathers from a craft shop. Examine the feathers. Cut a cross section of the shaft to expose the hollow section. (3-6)

- 14. Check down coats or pillows to find soft down feathers. Compare with other types of feathers. Use a magnifying glass. (3-6)

- 15. On a cold winter night, go outside in your pajamas to feel the coldness. Discover how down feathers keep birds warm, by wrapping yourself in a **down comforter**. (3-6)

- 16. Use pictures to get a closer look at **flight feathers**. (3-6)

- 17. Brainstorm some possible uses for feathers. Feathers have had many historical uses: pen, fan, jewelry, arrow, fishing lure, decorations, hats, and on clothing. (4-6)

- 18. Write with a **feather pen**. (5-6)

- 19. Look at Native American feather headdress and feather jewelry. (3-6)

- 20. Go on a, **"feather hike."** See who can find the most. Always wash your hands after any nature hunt. (2-6)

- 21. Use a magnifying glass to examine the skin of a whole, uncooked turkey's wing. Look for holes where feathers were attached. Try inserting a craft feather into the hole. (4-6) Wash your hands afterwards. Use clay to make birds. Decorate by sticking feathers into the clay.

- 22. . Each year the **Arctic tern** migrates thousands of miles, farther than any other bird. Learn more about these birds. Use your globe to determine their migration path. (4-6)

- 23. The **Emperor penguin** lives in the South Pole. These birds are about the height of a 5-year-old child. Learn about these fascinating birds. Discover places other than the South Pole where penguins live. (3-6)
- 24. Pretend to be an Arctic turn migrating. How will you find your way across land and water as you migrate? (3-6)

- 25. The **snowy owl, snow goose,** and the **ptarmigan** live in the Arctic tundra. During the winter months when the tundra is frozen the bird's feathers turn white. This coloring helps to camouflage against the white snow. Discover more about these birds as well as other Arctic animals whose coats turn white or who are white.-polar bears, arctic foxes, Arctic hare, ermines, weasels, beluga whale (3-6)

- 26. Learn about other birds that live in the North and South Poles. (3-6)

- 27. During the winter food is scarce. Make **bird feeders**. Attach a string to a pinecone. Cover it with peanut butter and roll in birdseeds. Hang it where birds can reach it. (2-6)

- 28. Make sure that the food is placed where birds can reach it. A good problem-solving activity is to figure out ways to position the food, to keep other animals from getting to it. (4-6)

- 29. String popped popcorn and hang it in a tree. (2-6)

- 30. Other suggested foods include breadcrumbs, broken dog biscuits, dried fruits, and peanuts. (3-4)

- 31. Supermarkets, garden and pet stores have prepared bird seed cakes that are blended and ready to hang. Research the Internet to find recipes for making different kinds of bird feed mixtures. (2-6)

- 32. Birds also help in the **recycling** of the earth's dead plants and animals. **Vultures** consume the dead and decaying bodies of animals. Learn more about these and other birds that are **scavengers**. (4-6)

- 33. Build or buy a **birdhouse**. Most craft stores carry wooden birdhouses that are ready to be painted. By having a birdhouse in your backyard, you will encourage birds to nest there. Make sure that you place the birdhouse high enough that other pets cannot disrupt it, but low enough that you can lift your child to peek into check on eggs and baby birds!
- 34. Discover birds that are **endangered**. Find a picture of the **dodo bird**. Many years ago, these birds were hunted to **extinction**. (4-6)

- 35. The island of Guam in the South Pacific Ocean has very few birds. The brown **tree snake** (not a native snake) was introduced to the island. In a few short years, these snakes all but wiped out the island's bird population. There is now a movement to rid the island of these snakes and to restock it with birds. Use your globe to find out where Guam is located. (4-6)

- 36. In the spring, it is not uncommon to find an injured baby bird that has fallen from its nest. Many areas have animal rescue centers where you can take injured birds to receive medical attention. Caution about not touching wild animals. (3-4)

- 37. Examine a chicken's egg. Discuss the shape. Crack it open and wash the inside of the shell. Let your child handle the shell. Examine the shell's contents. The yellow is the yolk and if fertilized and kept under ideal conditions, would develop into a bird. (2-6)

- 38. Caution your child about eating **raw eggs**. Although most eggs are safe, some may contain **salmonella**-sometimes-deadly bacteria carried in raw eggs. Eggs must be cooked to kill the bacteria. Some turtles also carry the salmonella bacteria. Always wash hands after handling raw eggs. (3-6)

- 39. An **ostrich** is the world's largest bird. It can grow to approximately 8 feet tall. Learn more about these birds. Measure 8 feet. Use a globe to discover where ostriches are found. (2-6) Find things around the house that are 8 feet tall. Compare an ostrich's height to the adults and children living in your house.

- 40. An **ostrich** egg is approximately 8 inches long. Make a papier-mâché ostrich eggs. Inflate a small round balloon to approximately 8 inches long and tie. Prepare a glue mixture (1 part liquid white glue and 3 parts water), dip strips of newspaper into the glue and apply to the balloon. Cover the balloon with 3 or 4 layers of newspaper. Let each layer dry, before applying another. Paint the egg with white paint. (3-6)
- 41. Compare the size of an ostrich's egg to a chicken's egg. Many stores have duck and quail eggs. Compare the eggs to duck and quail eggs, inside and out. (3-6)

- 42. Research the **smallest** bird, the **hummingbird**. The smallest hummingbird-the bumblebee hummingbird is 3 inches long. Sketch and cut-out a bird that is 3 inches long. Use the bird to find things that are approximately 3 inches. Compare the size of an ostrich (8 feet tall) to a hummingbird. (3-6)

- 43. Cook or buy a **whole chicken** for dinner. Carefully remove the meat. Try to leave the carcass (bones) intact. Examine the bones. Compare the size and shape to beef or pork bones. (2-6) How are they alike? Different?

- 44. Use books to discover some of the many different kinds of **feet** and **bills** of birds. (4-6) Research and compare how they use their feet and bills. Make sketches. (4-6)

- 45. Many stores sell chicken's feet. Examine the feet. How are they like your feet? Different? Boil the feet and remove the meat (try to keep the bones intact. Also buy whole pig's feet. Compare these bones to those

Mary Taylor Overton

of the chicken. Look at pictures of the feet of other birds-ducks, owls, ostriches. Compare the feet of other animals (4-6)

- 46. Discover how owls use their asymmetrical ear placement for hunting. (3½-6) Pretend to be an owl sitting in a tree listening for a mouse running around on forest floor. Use your asymmetrically placed ears to identify the direction that the mouse is running. Owls swallow their food whole and later spit out the bones.

- 47. What is the name of your **state's bird**? Learn about it. (4-6) Find out the names of birds of the states touching your state. Is there a sports team in your state that is named for a bird?

- 48. Our national symbol is the **American Bald Eagle**. Learn about this bird. A baby eagle is called an **eaglet.** (3-6)
➤ Discover the many places where its picture can be found.
➤ Use a string to measure to show the size of its nest. An eagle's nest can be over 5 feet across, about the size of a full sized bed.
➤ Why was it once endangered? How was it saved?

- 49. Pretend to be a bird. (2-6)
➤ Flap your wings and fly
➤ Glide
➤ Swoop
➤ Land
➤ Soar
➤ Hover like a hummingbird
➤ Dive for food
➤ Walk
➤ Hop
➤ Sit on a nest containing eggs
➤ Balance and perch on a branch
➤ Use your bill to make a nest
➤ Sing
➤ Fly against the wind
➤ Migrate
➤ Swim

- 50. Read about birds that are excellent underwater swimmers-**penguins**. (3-6) Compare the wings of penguins to those of other birds.

- 51. Draw a bird. Help your child decide what kind of bird to draw. Look at pictures and talk about features. What kind of beak should it have? Feet? Feathers? Where does it live? Glue real feathers. (4-6)

- 52. Some birds like **pelicans** raise their young in a community unit with other pelicans. The adult birds act as baby sitters, keeping a watchful eye over all of the young birds. They make sure that no predators get too close. Learn about pelicans and other birds that live in a community group. (3-6)

- 53. Learn about the **peregrine falcon**. This bird is the fastest flyer of all. Discover other fast flying birds. (4-6)

- 54. Research birds that do not fly, they include **kiwis**, **ostriches**, and **rheas** ($3\frac{1}{2}$-6)

- 55. **Birds of prey** are often referred to as the **raptors**, this means, "to **seize and carry away**." They include **hawks, condors, falcons, owls, eagles**, and **stalks**. These birds help to control the rodent population. Find out more about these birds.(4-6)

Mary Taylor Overton

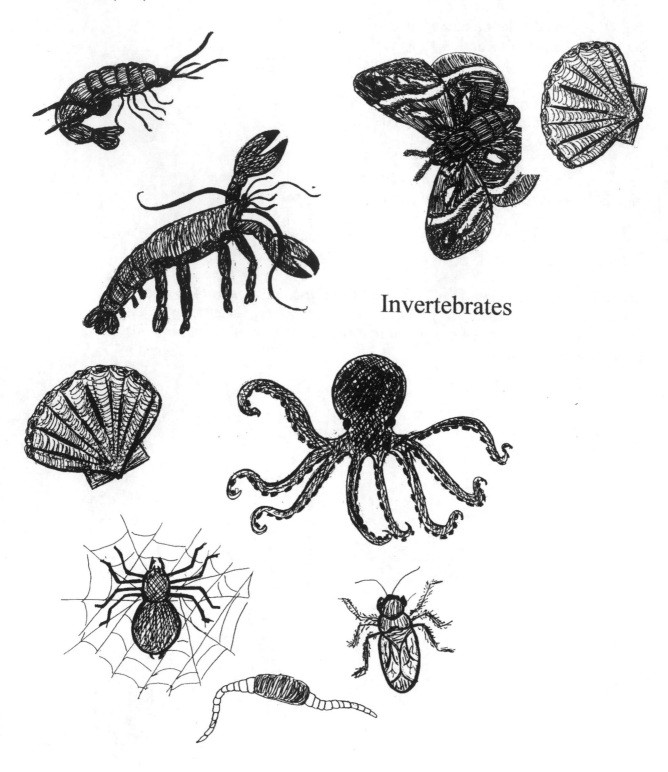

Invertebrates

Invertebrates(2-6)
Parent Background
- Invertebrates are the largest group of animals.
- The animals classified in this group are among the most diverse. They include **earthworms, spiders, insects, worms, starfish, snails, clams, slugs, flatworms, giant squids, coral, octopuses, shrimp, crabs, lobsters, fleas, leaches, roundworms, sponges, roaches,** and **jellyfish** are only a few of the many animals classified as invertebrates.
- Invertebrates are **cold-blooded.**
- **Invertebrates do not** have **backbones.**
- Some have **exoskeletons** (skeleton on the outside).
- They can be found all over the world, living on land as well as in water.
- Many have a hard outer shell. (**crabs, insects, spiders, lobster, shells**)
- **Squids** have a hard inner section.
- **Jellyfish** and the **octopus** are entirely composed of soft tissue.
- Invertebrates are important link in the earth's **flow of energy** or **the transfer of energy.**
- **Invertebrates** are the **primary consumers** of plants, as well as small animal **herbivores** (plant eaters).
- Small invertebrates are a rich food source for other small and many large carnivores (meat eaters).
- Invertebrates consume a wide variety of different foods.
- ✓ **Herbivores**-(Plant eaters- caterpillars)
- ✓ **Carnivores**-(Meat eaters- spiders)
- ✓ **Omnivores**-(Plant and meat eaters- sea urchins)
- ✓ **Parasites**- (Animals that live off of plants and animals)
- Many invertebrates are classified as **decomposers**. The earth is a closed ecosystem; its natural resources must be recycled over and over. Decomposers feed upon dead animals and plants breaking down the tissues. Through the chemical reactions set in motion by decomposers, the minerals and nutriments in the tissues of dead animals and plants will eventually be broken down and absorbed into the surrounding soil. Nearby plants will absorb the recycled minerals and nutriments and use them to produce new plants.
- Invertebrate decomposers include **earthworms, centipedes, millipedes, carpenter ants, beetles,** and **sow bugs**. These are only a few of the thousands of decomposers who are always hard at work.

Suggested Invertebrate Activities (2-6)

- 1. Read about **invertebrates** and **vertebrates**. (2-6)

- 2. From the fish market, purchase a **crab** and **shrimp** (with the head on) compare the two. How are they alike? Different? ($2\frac{1}{2}$-6)

- 3. Compare the outer hard shell of a clam and to a mussel. Describe how they feel and look. (2½-6)

- 4. Purchase a **lobster** to examine. Compare it to a **crab**. (3-6)

- 5. Read about **giant clams**. These animals can weigh as much as 400 lbs. Find an animal that weighs approximately the same weight. Gather different species of clams. Examine the inside. (3-6)

- 6. Learn about the **giant squid**. These squids are the longest invertebrates; they can grow tentacles that are over 50 feet long. Use a ruler to measure 50 feet. Think of things that are 50 feet long (approximate 5 large cars). (3-6)

- 7. The **jellyfish** and **Portuguese man-of-war** have powerful tentacles that can reach over 60 feet. Learn about these strange creatures. Measure 60 feet. (4-6) If you visit the beach look for jellyfish.

- 8. Purchase a **squid** and an **octopus** from the market. Compare the two. Pull out the hard inner part of the squid. (3-6)

- 9. **Slugs** can be found in your garden or damp places in the yard. Slugs are snails without shells. Find slugs to observe. They are **nocturnal** animals. (3-6)

- 10. Set-up an aquarium, include some **water snails**. These animals will help to keep the tank clean. (3-6)

- 11. To see how a snail moves, wet a piece of glass and let the snail crawl on the glass. Observe the snail from the underside. Compare water snails to slugs. How are they alike? Different? Draw a slug. Pretend to be a slug slithering along the ground on a wet evening. (3-6)

- 12. Look at pictures of **coral**. If possible view videos of live coral. (3-6)

- 13. View the movie **Nemo**. This movie is perfect for finding out about the **Great Barrier Reef**. (3-6)

- 14. Craft and hardware stores sell natural **sponges**, buy one to examine. Find out about these animals. (3-6)

- 15. If you have a dog or cat, he might be suffering from **fleas**. Fleas are carnivores (meat eaters), called **parasites**. Your dog or cat becomes the host upon which the fleas feed. Learn all about fleas. (3-6)

- 16. **Ticks** are also **parasites.** Research ticks and how they survive by living off of their host. Learn about diseases carried by ticks. (3-6)

- 17. Most of the food chain's **decomposers** are invertebrates. Learn more about these decomposers. (4-6)

- 18. Research **millipedes** and **centipedes.** (3-6)

Insects-(2-6)
Parent Background
- Insects account for the largest numbers of animals on earth.
- They are classified as invertebrates.
- Scientists have identified thousands of species and continue to discover hundreds more every day. It is believed that tens of thousands of species have not been identified.
- Many insects are **endangered** or **threatened**, and hundreds are already **extinct.**
- Insects play a vital role in the **ecosystem**; they are one of the most important links in the earth's **food chain**. Most insects are **herbivores** (plant eaters) and the primary consumers of plants. This makes them a rich food source for small **carnivores** (meat eaters).

Insects and Spiders
- Many people mistakenly classify **insects** and **spiders** in the same group.
- They are members of two entirely different groups.
- When helping your child to discover the differences between insects and spiders, keep in mind that, "insects are **insects** and spiders are **arachnids**, and they are not the same."

Insect   Spider

Insects	Spiders
3 body parts	2 body parts
6 legs	8 legs
Antennae	No antennae
Wings (Most)	No wings

Insects-Have 3 body parts
1. **Head** has eyes, mouth, antennae or feelers located on the top of the head.
✓ **Eyes**- 2 Compound eyes-composed of hundreds of tiny lens
✓ **Mouth**-Varies depending upon species; types of food eaten.
✓ **Antennae**-Feelers; detects odors, gather vital information

2. **Thorax**-The middle section of the body- 6 Legs; wings
✓ Legs- 6 jointed legs- 3 on each side
✓ Wings-Most insects have wings

3. **Abdomen** –Last section. Many insects breathe through spiracles located on the sides of the body.
• Insects are **cold-blooded.** Cold-blooded animals cannot regulate their body temperature. Their bodies are not able to generate heat. If the air temperature is 50 degrees their-inner core temperature will be the same.

• **Exoskeleton**-Body support on outside-
• **Invertebrates**- No backbones
• To escape cold weather conditions, many insects **hibernate**. They remain dormant and tucked away in a warm spot until the spring.
• Some insects **migrate** to warmer places. The Monarch butterfly is such an insect; it travels thousands of miles south to warmer weather.
• In the fall of the year, many insects lay eggs and die; others spin a cocoon around their body and remain snug and warm throughout the winter.
• All insects **lay eggs**.
• Some insects emerge from their eggs, looking just like their parents.
• Others like **butterflies, ladybugs,** and **mealworms** are born looking nothing like their parents. They must go through several stages of development and changes before resembling their parents. This is called **complete metamorphosis.**
• Many insects go through an **incomplete metamorphosis**. They include **dragonflies, crickets**, and **grasshoppers**. These animals emerge from the egg in more advanced stages of development. It takes fewer changes to reach maturity.
• Insects eat a variety of different foods.
➢ Most insects are **herbivores**-plant eaters.
➢ Some are **carnivores**-meat eaters.

> ➤ Some are **insectivores**-insect eaters
> ➤ Many are **omnivores**-meat and plant eaters.
- Various patterns on each species body and the lifestyles of an insect provide **camouflage** (protection).
- Insects defend themselves in a variety of different ways.
- Some insects are **poisonous**.
- Most insects live **solitary** lives. Bees, ants, and wasps live in **cooperative communities**.

Vocabulary-

- Insects
- Head
- Thorax-The middle section - legs are attached.
- Abdomen- Last body part
- Antenna or feelers
- Legs-6 segmented legs
- Eyes
- Mouth
- Emerge
- Metamorphosis- To change
- Invertebrates-Animals without backbones
- Exoskeleton-Hard part of the body on the outside
- Cold-blooded
- Hibernate
- Prey

Cricket Observation-

Crickets are available from pet stores, for about 10 cents each. Put some sand or soil in the bottom of a clear plastic jar. Place the crickets in the jar, along with a small piece of an egg carton. Add small sticks for climbing, some cracker crumbs and a small medicine container cap turned upside-down to hold water. Cover the top of the jar with a piece of nylon netting (an old pantyhose will do). Secure with a large rubber band; your child is now ready to make discoveries about **crickets**.
(2-6)

Provide a **magnifying glass** for observing.
(2-6) Allow time to just observe the crickets. Give your child time to make discoveries on his/her own.

OBJECTIVES
1. To observe cricket
2. To identify body parts
3. To observe how crickets use their body parts

Mary Taylor Overton

4. To note similarities and differences to other animals
5. To observe crickets in their habitat

Suggested Questions to Help in Observing

- What do you see? (2-6)

- Do they all look alike? (3½-6)
-
- What color are they? (3-6)

- Are they all the same size? (4-6)
-
- Compare a cricket to yourself? How are they like you? Different?
- What do they have that you do not have? (3½-6)
-
- How are crickets like/different butterflies? (3½-6)
-
- Observe the crickets for a while to note what the activities they are engaged in. Come back later to compare any changes. Are they doing anything different? What are they doing now? (4-6)
-
- How many body parts do you see? (4-6)

- Can you see the eyes? (4-6) Learn how insects see. Find out what compound eyes are?

- How many legs do you see? Try catching one of the crickets. Observe how it uses its hind legs. Can you jump like a cricket? (4-6) Make sketches of the cricket's legs. Are all of his legs alike? Compare its hind legs to the front legs. How are they alike? Different?

Suggested Cricket Activities-(2-6)

- 1. Read books about crickets. (2-6)

- 2. Discover animals that eat crickets? (3-6)

- 3. You may have crickets in your basement, try catching one. These crickets do not look like the commercially raised crickets found in pet stores. (3-6) Compare the two.

- 4. On a warm summer night listen to the **sounds of crickets chirping**. (2-6)

- 5. Discover how crickets make their chirping sound? (3-6)

- 6. Research to find other insects that make a noise-
  (cicadas) (4-6)

Other Insects to Observe
-Ants   -Butterflies and moths
-Roaches   -Mealworms
-Bees   -Grasshoppers
-Ladybugs   -Beetles

- **Ants**-Ants are social animals. They live and work together in colonies. A single queen ant produces all of the eggs for the colony. Each occupant has a specific job to perform. If you would like to observe ants over a period of time, set-up an ant colony. Many toy stores also stock "ant kits." These kits have instructions for ordering live ants. You can also observe ants by putting a small piece of fruit on the ground to attract ants. After about an hour. Sit and watch what the ants are doing. Follow the ants to see where they are going.

- **Mealworms** look like worms, but are actually insects that have not gone through a complete metamorphosis. They may be purchased from the pet store for about 10 cents each. These insects go through a series of easily identified **metamorphic** stages. When you purchase mealworms they will look like worms. This is the **larvae** stage of a black hardback beetle. Store the mealworms in a large flat plastic container with a layer of cornmeal. For moisture add a small piece of an apple. Change the apple and check the water often. Every few days check the mealworms to note change. Below are the

- Stage you should see- mealworm's **cycle- egg, larvae, pupae,** and **hatch into black hardback beetle.** These beetles cannot fly. The adult female beetles mate and lays eggs, and the mealworm's cycle begins again.

(This project is one that requires time to complete the **cycle**, but it is worth the effort)

Keep checking to observe the different stages of development.

Butterflies and Moths-(2-6)
Parent Background
- **Butterflies** and **moths** are **insects**.
- **Butterflies** are active during the daytime, while **moths** are usually active

during the evening and nighttime hours.
- Contrary to most beliefs, moths are not always brown and ugly.
- Butterfly hatching kits may be purchased at garden and toy stores.

Vocabulary
- **Egg**-Contains the embryo
- **Larva**-When a caterpillar emerges from the egg
- **Pupae**-State of changing inside of a **chrysalises or cocoon,** while changing into a butterfly or a moth
- **Chrysalises**-A protective casing spun by a butterfly
- **Cocoon**-A protective casing spun by a moth and other insects
- **Molt**-Insect's skin cannot grow. As they grow the skin splits to reveal the new larger skin.
- **Emerge**-To come out of the **chrysalises, cocoon** or **egg**
- **Metamorphosis**-To change from a caterpillar to a butterfly or a moth.
- **Proboscis**-A straw like mouth, used to suck nectar
- **Nectar**-Sweet food made by **flowers**
- **Migrate**-To travel from one area to another
- **Four wings**-Attached to the **thorax**

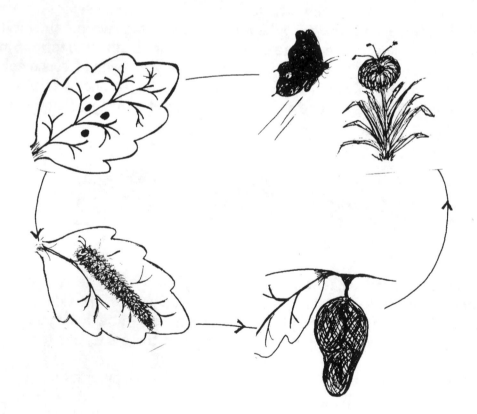

The **life cycles** begin when a female adult butterfly or moth lays eggs on the leaves of a plant. The eggs hatch and small **caterpillars (larvae)** emerge. Right away the hungry caterpillars eat the egg casings and then the leaves of the host plant. The skin of caterpillars does not grow, so when it gets too small it splits (**molt**), revealing a new larger skin. The caterpillars will **molt** several more times, before moving to the next stage of development.

Caterpillar (larvae)

Next the caterpillars begin to **pupate**. They hang upside-down, attaching to a twig and spinning a **chrysalis-butterflies** or **cocoon–moths** around their body. While in the pupae stage the caterpillars go through a complete **metamorphosis**–When they emerge from the pupating stage their appearance will have changed completely. They have **3 body parts**, **4 wings**, **6 legs**, and **a pair of antennae**. This is the adult stage of moths and butterflies.
- The wings are covered with colorful scales with matching **symmetrical** patterns on each wing.
- Butterflies and moths have a **proboscis** for feeding. This is a long hollow straw-like curled tube that is uncurled for sucking sweet nectar from flowers.

Butterfly, Moth and Insects Activities (2-6)
- 1. Read books about butterflies, moths, and caterpillars. (2-6)

- 2. To attract butterflies, plant a butterfly flower garden. Prepared butterfly flower mixes are available at garden stores.

- 3. Watch butterflies as they fly around. Observe their behavior. (2-6)

- 4. In the spring of the year look for caterpillars. (2-6)

- 5. Xerox blank caterpillars for your child to create his/her own species.

- 6. Copy and cut apart the **Butterfly Life Cycle cards** at the end of the unit. Use the cards to tell the story of the cycle of a moth or a butterfly. (2-6)

- 7. Butterflies and moths go through a complete metamorphosis. Research other animals that go through a complete **metamorphosis**? (frogs, flies, salamanders) (2-6)

- 8. Learn about **incomplete metamorphosis**. (5-6)

- 9. Pretend to be a caterpillar eating leaves and getting fatter. You get so fat that you must molt. **Molt** by splitting your skin to show-off your new skin. ($3\frac{1}{2}$-6)

- 8. Compare a moth to a butterfly. (4-6)

- 9. Compare a butterfly to an ant. (3-6)

- 10. Use a party blower (the kind that rolls and unrolls when you blow into it), to imitate the butterfly's **proboscis** (mouthpiece). Pretend to be a butterfly landing on a flower. Unroll your proboscis to suck the nectar from the flower. (3-6)

- 11. Pretend to be a butterfly sucking nectar, use a straw for a proboscis. Drink your nectar (juice) from a flower. ($3\frac{1}{2}$-6) Discover other animals that suck nectar, such as **hummingbirds** and **bees**.

- 12. Learn about the **monarch** butterfly. These are migratory insects. Use the globe to trace their route. (3-6)

- 13. Discover other insects and animals that migrate. (4-6)

- 14. Find out what species of butterflies can be found in your area. (4-6)

- 15. Read books about the **Rain forest**, there are many unusual insects living here. Look at a globe to discover where the rainforests are located. (4-6)

- 16. Many insects like mosquitoes are harmful to people. Find out about these creatures. (2-6)

- 17. Catch a housefly; use a magnifying glass to get a closer look. Caution your child about not handling flies. Theses little creatures often carry diseases. (2-6)

- 18. Catch **fireflies**, on a warm summer night. (2-6)

- 19. On warm summer evenings, a porch light will attract **moths**. Watch to see what they are doing. Carefully catch a moth. Use a magnifying glass to note body parts and make sketches. Always release the moth. (2-6)

- 20. Learn about anteaters and other animals that eat insects. (2-6)

- 21. Visit a stream or a pond to observe some of the insects on or near the water. (2-6)

- 22. Live ladybugs are available from any garden store. Observe. Discover other beetles. (2-6)

- 23. Visit a field to find insects living there. (2-6)

- 24. Discover the largest insect. (2-6)

Spiders -Arachnids(2-6)
Parent Background
- **Spiders** are a member of a group of animals classified as **arachnids**. They include **spiders, scorpions, ticks**, and **mites**.
- They are classified as **invertebrates**.
- Spiders are one of the most misunderstood creatures living on earth. There have been so many scary stories and myths written about these little animals, just the mention of spiders causes panic.
- Contrary to beliefs, **most** spiders are not poisonous to people and only a few species are actually harmful to humans.
- Everyday spiders eat millions of insects. Without spiders, insects would overrun our world.
- Spiders are a vital link in our **ecosystem** and many of the earth's food chains.
  ### Characteristics of spiders
- Spiders are cold-blooded.
- All spiders are **carnivores**-meat eaters
- Many are cannibals. The black widow spider eats her mate.
- **Arachnids** have **Exoskeleton**- Hard outer shell.
- **Invertebrates**-No backbone
- **Spiders** grow by **molting**
- 2 main body parts- (1) **Cephalothorax**-means head and chest are combined into one-piece
- (2) **abdomen**-8 jointed legs attached to the cephalothorax

181

- Its body is covered with tiny hairs.
- **Spinnerets**-2 small holes located at the end of the abdomen. Silk for spinning a web flows from the spinnerets in a liquid form and hardens when the air hits it.
- Most spiders weave webs. There are many different kinds of webs. Webs are very strong and able to catch animals many times larger than the spider. In terms of holding strength, a spider's web has been compared to lightweight steel.
- Web makers wait patiently until an animal gets tangled in their web.
- **Jumping spiders** attack by pouncing upon their prey.
- **Trapdoor spiders'** ambush their prey by hiding underground, concealed by a flap over the trap's hole.
- Hairs located on the spider's body can detect vibrations from its surroundings.
- Whether they spin webs or not, all spiders depend upon the element of surprise. Most sit and wait for dinner to come along.
- Spiders have **8 eyes**.
- **Palps**-Sensory feelers are located on the front of the head.
- Spiders do not have teeth.
- **Fangs** are located on the front of the head. Spiders use their hollow fangs for injecting their prey with a poison that paralyzes or kills. The injection turns the prey's insides into a liquid. The spider then sucks the liquid, leaving the prey's outer shell.
- Spiders lay their eggs in an **egg casing.**
- Baby spiders are called **spiderlings.**
- Most do not care for their young.
- **Dragline** is a thin piece of silk spun by spiders and used to travel quickly from high places, such as from the ceiling to the floor.

Suggested Spider Activities (2-6)

- 1. Read about spiders. (2-6)
- 2. Discover how spiders and insects are different? Alike? (4-6)

- 3. Compare a spider to a dog. (4-6)

- 4. Make spiders from clay. (4-6)

- 5. Look for spider webs in your house. (2-6)

- 6. Observe a spider in your yard. (2-6)

- 7. Find pictures that show step-by-step how spiders spin webs. (4-6)

- 8. Research the different types of webs made by spiders. (4-6)

- 9. Find out about **poisonous** spiders. Which spiders are harmful to humans? (4-6)

- 10. Discover the largest spider- **The bird eating spider**. (4-6)

- 11. Read about the **wolf spider.** This spider carries its babies on its back. (4-6)

- 12. Discover the **jumping** and the **trap-door spiders**.(4-6)

- 13. Discover how spiders **camouflage** themselves. (4-6)

- 14. Research the **crab spider**. This spider is able to change its colors to blend with its surroundings. (4-6)

- 15. All spiders are **carnivores**; find out the many different kinds of prey eaten by these little creatures. (4-6)

- 16. Discover animals that prey upon spiders. (**Frogs**, other **spiders, praying insects, birds, skunks, turtles, moles, shrews**) (4-6)

- 17. Pretend to be a spider waiting for a meal to come along. Decide the kind of prey you will be having for dinner. How you will capture it? Describe the process you will use for eating and digesting you dinner. ($3\frac{1}{2}$-6)

Worms-Earthworms(2-6)
Objectives-
1. To identify earthworms
2. To learn how earthworms fit into the food chain
3. To observe the earthworm's body
4. To understand how important earthworms are to our environment
5. To identify where earthworms are found

Vocabulary-
✓ Earthworms
✓ Soil
✓ Rings or segments

Parent Background-
- The **earthworm** is an **invertebrate**; it does not have bones.
- They are **cold blooded** and cannot regulate their body temperature.
- Earthworms do not have eyes or ears.
- They live underground.
- Their bodies are sensitive to light, cold, and heat.

- The body must be kept moist.
- They breathe through their skin.
- Earthworm's bodies are long and soft with **rings** or circular segments encircling.
- Prolonged exposure to air will cause their skin to dry out.

- They are **nocturnal** animals, emerging at night to hunt for rotting and decaying plants and animals.
- There are millions of earthworms living in the dark moist soil. The earthworm's diet makes it a good food source for many small carnivores and omnivores eaters.
- Earthworms are classified as decomposers. They eat decaying and rotting plants and animals.
- Earthworms perform an invaluable service in the earth's food chain. After digestion is completed the excreted waste (containing minerals and vitamins) is deposited and absorbed into the surrounding soil. These minerals will be eventually recycled by nearby plants.
- Earthworms are often referred to as, "**nature's fertilizer makers**."
- They also create millions of underground tunnels systems that allow air and water to get down into the soil. The roots of growing trees and plants must have oxygen to perform **photosynthesis** (to make food).

Suggested Activities-Earthworms (2-6)

- 1. Earthworms may be purchased from pet stores. They are also found in your garden. Wet a small section of soft earth and after dark; use a flashlight and a small shovel to gently dig into the soil to locate earthworms. Store the earthworms in a large jar that has been filled with soil that you have collected near the worms. Place an apple peels on top of the soil.
(Use a magnifying glass to observe) (2-6)

- 2. Describe the earthworm?

- 3. What happens when it is exposed to light?

- 4. Which end is the head? The tail? (2-6)

- 5. How does it feel? (Use descriptive words- soft, cold, wiggle, and etc.)

- 6. What happens when you sprinkle the earthworm with water?

- 7. Can you see the rings on its body? (2-6)

- 8. How does the earthworm move? (2-6)

- 9. Compare earthworms to other animals.

- 10. Discover other kinds of worms. (2-6)

- 11. Earthworms are **cold-blooded** (cannot generate its own body heat) and have difficulty moving in very cold weather. However, they can be safely stored in the refrigerator (NOT THE FREEZER) to be kept for further study. Put a few earthworms into the refrigerator and allow them to get cold. What did you discover?

Butterfly or Moth Cycle cards

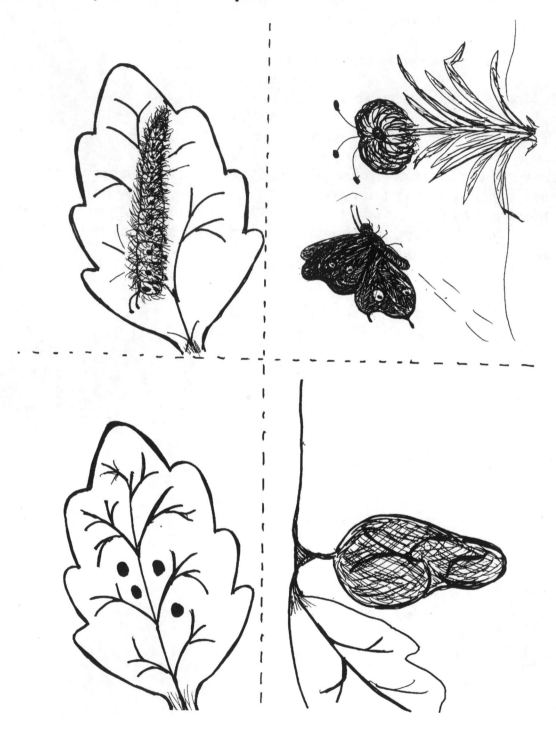

The Sun (2-6)
Parent Background

- Nothing impacts or influences our lives more than the **sun**.
- It directly and/or indirectly controls every aspect of all living and most non-living things on earth.
- Without the sun, life on earth as we know it could not exist. At a very early age, children should begin to understand how connected their lives are to the sun. The sun is a source of wonder for children. It is always in sight and can offer tons of math and science experiences. It's free, so get out and observe and discover the sun and its properties.
- The sun is a **star**, composed of hot gasses.
- The **universe** is filled with thousands of stars (suns); some may even have planets with life that is similar to life here on earth.
- Our earth is one of nine planets in our **solar system** with our sun its center.
- All nine planets orbit around the sun. It is the **orbital** position of each planet that determines the amount of heat and light that each receives.
- The **earth** is the third planet from the sun. Its orbital position and its natural resources are just right for supporting life.
- It takes the earth 365 days (1 year) to completely circle the sun.
- The sun plays an important role in our lives. It supplies all of the earth's natural **sunlight**. All living things on earth are dependent upon the sun for energy.
- Earth's green plants are the direct **receiver** and **dispenser** of the sun's energy.
- **Chlorophyll**, a green substance in the plant's leaves, absorbs the sun's light **energy**. Using the energy for power plants produce their own food. Plants use the energy packed food to carry out their basic needs.
- Unused energy packed food is stored within the plant's **flowers**, **fruits**, **seeds**, **leaves**, **stems**, **pods**, and **roots**. Animals and people eat these plants as food.
- Their bodies then burn the stored energy in these plants to maintain their daily activities. This is known as the **food chain** or **transfer of energy**.
- The earth's round spherical shape causes an uneven heating from the sun's rays striking the earth's surface. This results in equator receiving more direct heat than any other place on earth.
- The hot air from the equator spreads out across the earth. As the warm air moves it, comes into contact with a variety of air temperatures and atmospheric pressures. This results in different weather patterns across the earth. The earth's air is constantly moving.
- As the earth travels around the sun, it also rotates on its **axis**. The earth is slowly turning, so that only $\frac{1}{2}$ of the earth is facing the sun at any one time. This rotating results in the earth having **day** and **night**. It takes 24 hours for a complete rotation. While your side of the earth is having day, the other $\frac{1}{2}$ of the earth is in darkness.

Mary Taylor Overton

Objectives-
1. To understand the properties of the sun
2. To develop related vocabulary
2. To observe the sun
3   To understand the connection between the sun and the earth
4. To gain a healthy respect for the powerful effects (both positive and negative) that the sun can have on people, plants and animals
5. To note seasonal changes
6. To explore day and night

Vocabulary

- Sun
- Rays
- Sunlight
- Shadow
- Sunrise
- Sunset
- Reflection
- Solar system

- Heat
- Food chain
- Sunburn
- Cold
- Hot
- Seasons
- Winter
- Summer

- Spring
- Fall
- Temperature
- Weather
- Planet

Suggested Sun Activities (2-6)

- 1. The sun is a **sphere**. Explore the properties of spheres. (See sphere activities in the geometry section) (2-6)

- 2. Discover some of the properties of the sun. (3-6)

- 3. Read books showing the sun's and earth's positions in our solar system. ($3\frac{1}{2}$-6)

- 4. Young children enjoy learning how large the sun is in comparison to the earth. On a piece of yellow construction paper, trace a circle around the edge of a large dinner plate and cut it out. This will represent the sun; use a green pea to represent the earth. Compare the sizes of the two. (4-6)

- 5. Purchase an inflatable globe. They are available at any dollar store. They are excellent for locating the weather that is coming your way. You and your child should view the TV weather station to check approaching weather. Use the globe to show the weather's path. The globe is also excellent for locating the Arctic or North Pole, South Pole, Antarctica, the Rainforest, tropical weather, the direction of hurricanes, learn where the crab that he is eating comes from, locating migration routes of animals, discovering where different species of animals live, and a whole host of other math and science experiences. (3-6)

- 6. Make **sunny day** pictures. (3-6)

- 7. **The need for sunscreen**. The earth's atmosphere acts as a giant sunscreen by blocking the sun's harmful rays. Humans need further protection from the sun. What do you think would happen to your skin if you remained in the hot sun too long? Discuss **sunburn** and the need for wearing protective sunscreen lotion. Animals like pigs and elephants can get sunburned too. However, they cannot go to the store to buy sunscreen lotion. What do you think they can do to protect their skin? (mud) (2-6)

- 8. Place an outdoor **thermometer** in sight and use it for observing how the mercury rises and drop. Discussing the use of the thermometer should help your child make the connection that it is a measuring tool and how to use it in his/her life. Use it to compare how the **mercury** is up when it is **hot** outside and down when it is **cold**. Check the thermometer in the early afternoon and late in the evening, this is when the greatest contrast should occur. (4-6)

- 9. (4-6) Use 2 different color washable markers to indicate temperature levels. Use 1 color for the afternoon (The afternoon temperature is usually the warmest), and the other marker for the evening. Compare your findings. What can you tell me about where the temperature line is in the afternoon? Where is it in the evening? Use descriptive words such as (up, for hot, warm), (down, for cold, cool). "It is going to be a hot day; will the mercury be up or down?"

- 10. The **weatherman predicts** that the temperature will be 80 degrees today? Will the mercury be up or down? This is an on going project. As your child's math and science skills and knowledge grows he/she will be able to make more connections about the sun and the part that it plays on **animals, people, plants, weather**, and **seasonal changes**. (4-6)

- 11. Read about cold-blooded animals-**reptiles, amphibians**, fish, and **invertebrates**. These animals depend upon their surrounding to keep warm. (2-6)

- 12. When it is too cold, many animals **hibernate**. When it is too hot, many **cold-blooded** animals go into hot weather hibernation called **aestivating**. Research **hibernation** and **aestivation**. Discover some of the animals that engage in each. (3-6)

- 13. Put 2 dishes containing cubes in the sun. How can we one from **melting**? Put one into insulated cooler and set the cooler the sun. Leave the other in the What do you think will happen? (3½- ice keep an in sun. 6)

- 14. The sun provides **heat** to the earth. On a sunny day, select things to place in the sun. Choose 2 of each item. Put one set directly in the sun, and the other inside. What do you think will happen? After an hour check and compare the 2 sets. What did you discover? Try using things such as plastic, wood, rocks, paper, metal, and glass. (3-6)

- 15. Get up early to watch the **sun rise** over the **horizon**. Discuss how it looks outside before the sun comes up. Is it dark? Can you see the sun? From which direction do you see the sun rising? What **colors** do you see? How does it look outside, now that the sun is up? (3½-6) Observe the **sun set**. What direction is the sun **setting**? What colors do you see? Compare sunset to the sunrise? How does it look now that the sun has set? (2-6)

- 16. Where does the sun go when it **sets**? (2-6)

- 17. What causes **day** and **night**. (2-6)

- 18. Research animals that are active at night-**nocturnal**. (2-6)

- 19. Discover animals that are active during the day-**diurnal**. (2-6)

- 20. Look for days when you can see both the **sun** and the **moon** in the **sky** at the same time? (3-6)

- 21. When you see the sun in the sky, why does it look so small? Explore the properties of **distance** (3½-6).

- 22. Read books about the **North and South Poles**. The South Pole is the coldest place on earth; this is because it receives the least amount of direct sun's heat. Read about animals from these poles. Many fiction books show polar bears and penguins together; however, they live at opposite poles. Learn about each of these animals. (3-6)

- 23. Look for your **shadow** and the shadows of other things. What do you think is making the shadow? Try moving the shadow around. (2-6)

- 24. Many years ago people used sundials to keep track of time. Research and make a **sundial**. Find a pole or put a stick vertically into the ground. Mark the shadow's position. After 2 hours check the position of the shadow and the stick. What did you **discover**? Leave the marker and check again one hour later. What happened this time? (3½-6). Did the stick move?

- 25. Remove a lampshade or use a flashlight to project light on a wall. Use your hands to make funny moving shadows on the wall. Try using other objects. (2-6)

- 26. On a cloudy day try finding your shadow. (2-6)

- 27. Where is the sun on cloudy days? (2-6)

- 28. Read books about plants and how they use the sun's light. (2-6)

- 29. Visit a garden store to buy a houseplant to care for. (1½-6)

- 30. Plant seeds in 2 different containers. Place 1, so that it gets lots of sunlight and the other in a dark closet where it will not get any sunlight. What do you think will happen? Observe both plants to see what happens. This is a project that will take time to develop. (3-6)

Seasons-(2-6)
Parent Background-
- Our sun controls all life on earth.
- The earth is always tilted on its axis at a 23.5-degree position.
- This is an imaginary line that runs from the North Pole, through the opposite end of the earth. The top point of the axis is always pointed toward **Polaris** or the **North Star**. This point never changes even as the earth travels its orbital path around the sun.
- The earth is divided into 2 halves, the **northern** and **southern hemispheres.**
- The **equator** separates the 2 hemispheres.
- It is the earth's tilted position and its orbital path around the sun at different times of the year that determines the amount of heat and light each section receives. This results in weather changes from month to month for each hemisphere.
- The earth's spherical shape results in the bulging equator. This also results in this section of the earth receiving more of the sun's heat and

light.
- The opposite hemispheres have opposite weather occurring simultaneously.
- **Summer** occurs when one hemisphere is tilting towards the sun, giving this half of the earth more direct sun and heat.
- While this hemisphere is having hot summer weather, the opposite hemisphere is tilted away from the sun and receives less heat. This hemisphere is having **Winter.** Winter means a drop of temperature resulting in major changes for human, animal, and plants living in this hemisphere.
- **Fall** and **spring** are transitional seasons between winter and summer. As the tilted earth continues its orbit around the sun, it puts each hemisphere through transitional positions. This results in the milder temperatures of spring and fall.
- These changes result in the earth's **cycle of seasons**- 4 seasons- **summer, fall, winter,** and **spring.**
- Based on a seasonal weather changes, animals and plants react by adjusting their behavior to survive from one season to the next.

- Each season brings with it **predictable weather patterns.**

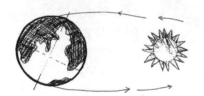 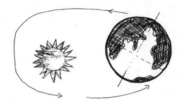

Objectives
1. To learn related vocabulary
2. To identify the 4 seasons
3. To learn that seasons follow a predictable sequential order
4. To understand that each season has its own characteristics
5. To understand that the behavior of people, animals, and plants changes with the season
6. To observe seasonal changes

Winter Vocabulary Building
- Weather
- Snow storms
- Cold, Colder, coldest
- Temperature
- Shorter day
- Warm clothing

- Sleet
- Blizzard
- Frozen
- Icy
- Slippery
- Frost on cars and lawns
- Frosty
- Icicles
- Snow drift
- Bare tree limbs
- Evergreen trees –Trees that stay green all year
- Conifer-Trees that make cones (Most are evergreen)
- **Animals Hibernate**- Many mammals rest or sleep during the winter months, when food is scarce. Reptiles, insects, and amphibians (cold-blooded animals) hibernate to keep from freezing to death.

Spring Vocabulary

- Weather
- Snow melts
- Warmer weather
- Rain showers
- Animals migration
- Animals come out of hibernation
- Leaves begin to bud
- Plants sprout (grow)
- Flowers blossom
- Buds on trees open
- Plant flower and vegetables
- Animal babies are born
- Warmer weather
- Days get longer
- Caterpillars

Summer Vocabulary

- Trees in full bloom
- Hot weather
- Animals raise their young
- Thunderstorms
- Drought
- Lots of flowers
- Fruit growing on trees
- Trees full of green leaves
- Butterflies
- Reptiles bask in the warm sun
- Lots of summer birds
- Insects
- Plenty of food for animals to eat

Fall Vocabulary-

- Days get shorter
- Autumn
- Harvest of crops
- Trees prepare for winter
- Animal's coats thicken
- Nippy weather
- Cooler weather

- Frost on cars and grass
- Dew on cars and grass
- Warmer clothing
- Red, yellow, orange, brown
- Animals prepare to **hibernate** (winter sleep
- **Migration**-Many birds and some insects travel to a warmer climate to find food and spend the winter.
- Deciduous trees-Trees whose leaves change colors

Suggested Seasons Activities (3-6)

Most of the following activities are suited for (3-6). However, 2½ will enjoy many of them, be selective. Take every opportunity to help your child to observe seasonal changes and to make connections about the sun.

Read books about:

➢ Seasons
➢ Cold-blooded animals
➢ Plants
➢ Trees
➢ Hibernation
➢ Migration
➢ The sun
➢ Flowers
➢ The food chain
➢ **Nocturnal animals**- animals that is active during the night hours.
➢ **Diurnal animals**- animals who are active during the day

- 1. Refer to the seasons in a **sequential order**: **summer**, fall, **winter, spring**. Make-up a chant. (3-6)

- 2. Take a picture of your child engaged in an activity during each of the seasons. Use the picture to show the **seasonal sequence**. (3-6)
➢ Can you find a picture of you in the **fall**?
➢ How do you know that it is **fall**?
➢ Show me a picture that shows the next **season**?
➢ Can you put the pictures in order? (3 ½ -6)

- 3. (4-5) Refer to, "Your birthday is May 18th, in the **spring**." "When your birthday comes, the trees will have new leaves."

- 4. Discuss and observe seasonal things: (2-6)
➢ Colors
➢ Trees
➢ Weather

- ➢ Sun's heat
- ➢ Leaves on trees change colors and drop off
- ➢ Clothing
- ➢ Animal behavior
- ➢ Human behavior
- ➢ Sports
- ➢ Holidays
- ➢ Foods
- 5. Compare and contrast the seasons. (4-6)

- 6. Draw pictures of things that occur each season. (4-6)

- 7. Discuss changes that people make to cope with each season. (3-6)

- 8. Select a tree in your yard to observe the changes that occur from season to season. Compare the seasonal changes. (2½-6)

- 9. Collect, compare, and sort leaves, seeds, acorns, and other by-products produced by plants and trees. (3-6)

- 10. Observe the animals that visit your yard at different times of the year.

- 11. During the winter months when food is scarce, feed the animals that visits your yard. Some suggestions are-(birdseeds, fruits, bread etc.). Attach a string to a pinecone, spread it with peanut butter and roll it in birdseeds. Hang the pinecone where birds can reach it. (2-6)

- 12. Observe a storm as it is happening. Talk about what you see? How does it feel? Compare rain to snow? Storms during hot weather can result in thunderstorms. Note changes in the sky and weather before, during, and after a thunderstorm. (3½-6)

- 13. **Eat seasonal foods**. We live in an age where most foods are available all year around. However, certain foods are more plentiful during each season. Brainstorm foods that are plentiful during the season the season that you are experiencing now. (3½-6)

- 14. Ask family and friends to name their favorite season. Make a graph to show the results. (4-6)

Weather Words

Mary Taylor Overton

Parent Background

Weather and its varying properties are concrete experiences that children can see, feel, hear and sometimes smell. Changes in weather can occur from minute to minute, hour to hour, day to day, week to week, month to month. There are always opportunities to use rich descriptive **weather related vocabulary.**

Weather Words:

- Meteorologist
- Forecast
- Weather station
- Temperature
- Thermometer
- Weather watch
- A clear day
- Sunny
- Partly sun
- Mostly sunny
- Drought
- Sunshine
- Warm
- Hot
- Humid
- Scorching sun
- Indian summer
- Spring like weather
- Winter like weather
- Summer like weather
- Fall like weather
- Windy

- Breezy
- Light winds
- Gusty
- Strong winds
- Steady rain
- Rainy
- Showers
- Sprinkle
- Rain showers
- Scattered showers
- Soaking rain
- Soggy weather
- Severe weather
- Light rain
- Heavy rain
- Cloudy
- Partly cloudy
- Mostly cloudy
- Thunderstorm
- Lightening
- Lighting strike
- Lightening bolt
- Rain clouds
- Rainfall

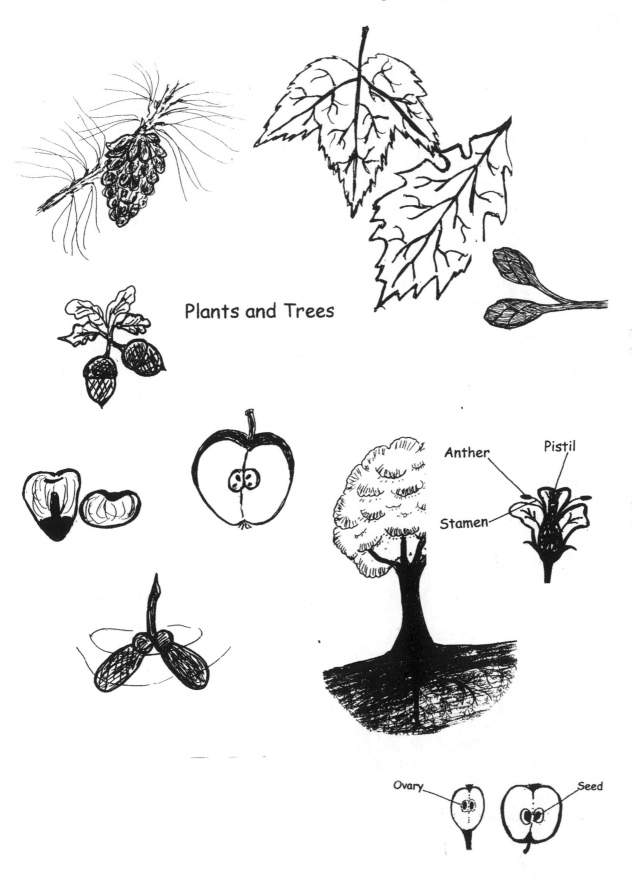

Plants and Trees

Anther

Pistil

Stamen

Ovary

Seed

Plants(2-6)
Parent Background
- Plants have existed on earth for millions of years.
- They are found on land, under water, in hot and cold climates.
- All living things on earth must have the energy that comes from the sun.
- The green plants are the earth's link to the sun's energy.
- Plants are the direct **receiver** and **dispenser** of the **sun's energy**.
- Earth's plants and animals are in a partnership with the sun, for their survival on earth.

Vocabulary
- **Blossoms**-Flower- seeds develop here
- **Branch**-Part of a tree
- **Buds**- Next year's flowers or leaves
- **Chlorophyll**- A green substance in leaves
- **Coniferous**-Trees that reproduce by cones
- **Cones**-Protective covering for developing seeds
- **Deciduous**-Plants whose leaves change colors and fall off in the fall
- **Dormant**- Sleeping
- **Embryo**-Baby plant within the seed
- **Flowers**-The part of the plant where fertilization takes place
- **Fruit**-Seeds develop inside.
- **Germinate**-The point at which seeds starts to grow
- **Glucose-**A sugar- starch produced by plants
- **Leaves**-The food maker
- **Limb**-Part of a tree
- **Nectar**-A sweet starchy watery food produced by flowers
- **Oxygen**-A by product released into the air by green plants as they make their food. Animals and plants must breathe oxygen.
- **Photosynthesis**-The process of making food powered by the sun's light energy.
- **Plants-**A non-animal living organism
- **Rings of a tree**-One ring develops each year in the bark of the tree
- **Roots**-Seek and absorb minerals and water from the ground
- **Root hairs**-Tiny thin roots
- **Sap**-A sweet watery food produced by trees
- **Seasons**-Summer, fall, winter, spring
- **Seeds**-Contain a baby plant and food
- **Seed coat**-A protective coating on seeds
- **Scatter the seeds**
- **Shrubs**-Woody plants that have more than one trunk
- **Sprouts**- A plant that is beginning to grow
- **Soil** –Dirt, earth
- **Stems**-The trunk of a flower-Carries water and minerals to all parts of the

plant
- **Sunlight**-Light energy from the sun used by plants to make food
- **Swell**-Seeds soak up water and begin to germinate
- **Tap root**-The main root
- **Trees**-A plant with a single woody stem
- **Trunk**-The woody stem part of a tree.
- **Twigs**-The smallest branch of a tree.
- **Veins**-Hollow tubes that allows materials to flow through the leaves
- **Waterproof**-Able to prevent water from entering
- **Weatherproof**-Able to keep out harsh weather conditions

Objectives
1. To learn the basic needs of plants
2. To learn how plants release oxygen into the air
3. To learn related vocabulary
4. To understand how to care for plants
5. To learn the role of plants in the transfer of energy- the food chain
6. To understand why plants are important to animals and humans

Plants and Trees
Plants (2-6)
Stems and Roots-
- The fleshy **stem** of a plant is the same as the woody **trunk** of a tree.
- **Root** and **stem** systems anchor and support the plant.
- Roots dig into the soil, seeking **water**, and **minerals** for the production of food.
- The root and stem systems are like hollow straws carrying water, minerals, and food to the plant's leaves, branches, roots, seeds, and flowers.

Leaves-
- Green leaves are the plant's food producers.
- Leaves are covered with small microscopic pinholes called **stomata**. Stomata open and close to allow the absorption and the release of the sun's light energy, water, oxygen, and carbon dioxide.
- Green leaves contain a green substance called **chlorophyll** that absorbs the sun's light energy. Powered by this energy, a chemical reaction occurs, resulting in the production of food. This is called **photosynthesis**
- Water, minerals, carbon dioxide (a by-product released into the air by animals and humans as they breathe) are mixed together to produce food. Plant food is a sweet watery **glucose** (sugar) food called **nectar** in plants and **sap** in trees.
- A by-product of photosynthesis is **oxygen**. Leaves release the oxygen into the air. **Oxygen** is a gas that humans and animals must have to sustain life.

- The plant's stem and root systems carry the energy-enriched food to all parts of the plant where it is used for growing and reproducing.
- Unused food is stored in the plant's stems, pods, roots, seeds, flowers, nuts, and roots.
- In the production of food, plants take in large amounts of water from the ground, through its roots. Large trees can drink 200 gallons of water in a day. Most of this water is not used and is released into the air through the stomata on the leaves.

Buds-

- Buds are the plant's future young leaves and flowers. Living inside of the tough outer bud shells are young flowers and leaves waiting for the right conditions to blossom. The shell is both water and weatherproof. This spring's buds were formed last spring. The young plants have remained tucked inside the buds all summer, fall, and winter. The warm spring weather will stimulate the buds to open and begin to grow.

FLOWERS-

- Many plants and most trees produce **flowers**.
- Flowers are usually brightly colored and/or may have a strong odor, to attract insects and birds to help in **fertilization (pollination)**.
- Most trees produce **flowers**. However, many of these flowers are dull and are often over looked.
- The seeds for the plant's future generations are **fertilized** and developed within the plant's flower.
- Most plants have both **male** and **female organs** within a single flower.
- **Pollen** is a powder (the male reproductive cells) found on the **anther.** The anther is located on the top of the **stamen** (the male sex organ). As insects seek the flower's sweet nectar bits of pollen get stuck to the body. While visiting another flower the pollen is transferred to the **pistil** (the female reproductive part of the plant) of this flower. This is called **cross-pollinating** or **fertilization.**
- Once pollination occurs, seeds begin to develop inside of the **ovary** (the female sex organ) of the flower.
- Some plants **self-pollinate** as the wind blows bits of pollen onto the pistil.
- The ovary grows larger and becomes the fruit of the plant. Growing safely inside of the fruit are the seeds for the plant's next generation.

Seeds-

- Plants reproduce by-**seeds, sprouts, bulbs,** or **spores.**
- The purpose of reproduction is to **preserve the species,** without it a species would die-out.
- Most plants reproduce by seeds.
- Seeds come in different colors, sizes, and shapes. Orchids produce the

smallest seeds, and coconuts the largest.

- Inside of each seed a small baby plant is waiting- **dormant**. The seed will remain there until the right **conditions** for growing are met.
- The **biological** make-up of a plant determines, how, when, and the number of seeds that will be produced. Some plants like peaches, plums, avocados, and mangos, develop one seed in each fruit. Oranges, apples, and pears, **produce** only a few seeds. Still other plants like green peppers, watermelons, cucumbers, dandelions, squash, and tomatoes produce many seeds.
- Each year, plants produce billions and billions of seeds. Only a small fraction of these seeds will ever develop into fully-grown plants.

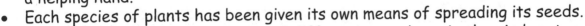

- If every seed grew to maturity plants would quickly over grow our world. Most seeds never find the right conditions for growing.
- To ensure the best chance for a seed to find an ideal location for growing, **Mother Nature** lends a helping hand.
- Each species of plants has been given its own means of spreading its seeds.
- Seeds are scattered in a variety of different ways-by animals, wind, water, self-dispersal, and people.
- Most plants **camouflage** their developing seeds until they are mature.
- While young oak tree seeds (acorns) are developing, they are green and camouflaged among the tree's green leaves. By autumn, when the seeds are mature, the acorns will turn brown. The brown coloring against the green leaves will stand-out attracting and possible encourage animals to pick the acorns. Perhaps some of the seeds will land in an ideal spot for sprouting.
- The seeds of an apple tree develop deep inside the apple's core. The apple's skin is green, until the seeds are fully developed. The outer skin will turn a vivid red color and will give off a fragrant smell. This fragrance is used to attract animals to eat the fruit.
- Some seeds are buried by squirrels or other animals and forgotten. If any of these seeds receive moisture and heat they will begin to germinate (grow).
- **Hitchhikers** are seeds that have sticky seeds coats that get caught on clothing and the coats of animals.
- Many seeds are swept away by rain or streams.
- Seeds are covered with a tough **protective** covering called a **seed coat**. It keeps the seed intact until conditions for sprouting are met. If a bird eats the fruit containing the seed, the seed's coat protects it as it travels safely through the bird's body. The whole seed will later be eliminated along with the animal's waste, far from the mother plant.
- Inside of the **seed coat** is a smaller version of the plant called an **embryo**. The seed also contains enough food to sustain the growing embryo until the **embryo** can grow leaves mature enough to produce food.

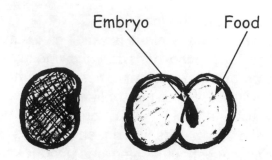

Embryo   Food

- Moisture will cause the seed coat to split. **Germination**
- Feeding off of the small amount of food stored inside of the seed, the **embryo** begins to grow.
- A **taproot** (this will be the plant's main root all other roots will grow from this root) grows downward into the soil. This root is looking for minerals and water that are needed by the plant for **photosynthesis** (making food).
- As the young **embryo** grows, it sprouts a **stem** upward seeking the warm light energy of the sun. Soon **leaves** will develop from the stem. The mature plant will produce seeds, and the cycle of the plant is complete. The new seeds will be used to carry on the next generation of this plant's species.

Trees (2-6)
Bark and Trunk-
- To support the massive weight of a tree, its stem is made of a hard woody material called the **trunk.**
- The hard outer layer of the trunk is called **bark,** and can be thought of as the **tree's skin.** It protects the tree from diseases, animals, and extreme heat and cold weather conditions.
- Each year, the tree's woody structure grows in thickness, pushing the hard outer bark out, leaving a visible **ring** throughout the tree's woody surfaces. By counting these rings, you can tell the approximate age of a tree.
- Trees are either **broad-leaf** (deciduous) or **conifers** (coniferous).

- **Deciduous** trees **Broad-leaved** trees have leaves that are **wide** and **flat.**
- These trees include-oak, maple, elm etc.
- They have thin delicate leaves that change colors and are shed during the fall season.
- ➤ Deciduous trees have a short food production and growing season of approximately 4-5 months out of the year. Therefore, they must produce as much food as possible during their growing season-**spring** and **summer.**
- ➤ The shape of deciduous trees, and the wide surface of their leaves, ensures that each leaf will receive maximum **exposure** to the sun's light energy.
- ➤ Each day trees must perform daily survival tasks. These tasks require a lot of energy.
- ➤ They get energy by producing their own food. (See the section on how

leaves.)

➤ For food production, there must be a constant flow of water from the tree's roots to its leaves.

➤ Freezing winter temperatures would make this almost impossible for the thin leaves of deciduous trees.

➤ To conserve energy during the winter season, **deciduous** trees stop making food.

➤ Any remaining **sap (food)** in the tree's parts is stored in the tree's roots until spring. The flow of water is stopped and the green chlorophyll color in the leaves begins to fade. The leaves now bear their fall colors and soon will drop to the ground.

• Once the leaves are on the ground animal and plant **decomposers** (mostly bacteria) attack the dead leaves. The job of the decomposers is to break down the remaining minerals and nutrients in the leaves, so that they can be absorbed back into the ground.

• All over the world millions of earthworms are busy underground working to decompose dead animals and plants. The earthworm will digest these materials and eliminate the minerals in its waste underground.

• Near by plants and trees will recycle these minerals and nutrients to make food in the spring.

• Next spring when the days become longer and warmer, **deciduous** trees will again become active. The food stored in the tree's roots will be sent up to the growing buds.

• This food will be used to sustain the trees until new green chlorophyll filled leaves can begin to produce food.

• There are a few broad-leaf trees whose leaves remain green all year around.

• Most **conifer** trees have leaves that are long thin and needle like. These trees are often called **evergreens**. They keep their leaves all year round, loosing only a few leaves at a time.

➤ The leaves of coniferous trees are covered with a thick waxy surface that protects them from the cold winter weather.

➤ Evergreen trees are able to produce food all year around. The shape of these trees also prevents snow from building-up on the branches and leaves.

➤ Many conifers are members of the fir, spruce, and pine tree families.

Bulbs

Some plants reproduce by **bulbs**. Bulbs are self-contained plants, much like the buds on a tree. Each bulb contains an embryo plant and a food supply. Roots sprout from the bottom of the bulb, a stem, leaves, and flowers will grow from the top.

Fungi

- Fungi are plants that do not contain chlorophyll.
- They cannot produce their own food.
- These plants attach themselves to other plants and animals and feed off their host's nutrients.
- Fungi include **mold**, **mushrooms**, **mildew**, **lichens**, and **truffles**.
- Many fungi live off of **decaying** animals and plants. They are actually helpful to the earth's foods chain. These plants breakdown the tissues of the plants and animals, to return their nutrients back into the soil to be recycled. These plants are called **decomposers**.

Suggested Plant and Tree Activities (2-6)
✓ **Most of these activities are appropriate for all ages, be selective.**

- 1. Read books about plants, flowers, and trees. Include a tree and flower identification manual. (2-6)

- 2. Learn the names of the trees in your backyard and near your home. (2-6)

- 3. Identify parts of a tree- trunk, branches, nuts, roots, twigs, and limbs of a tree. Examine the buds on the limb. (2-6)

- 4. The top of a tree is called the, "**crown**." Discover other things that have crowns-kings, the tops of bird's and people's heads, teeth, hats etc. (3-6)

- 5. Plant a tree. (2-6)

- 6. Compare the stem of a flower to a tree trunk. How are they alike? Different? (3-6) Examine the bark from several different trees.

- 7. Each year as a tree grows fatter; a visible ring is left on its internal structure. To show how the rings in the trunk of a tree are arranged, cut a large red onion in half, horizontally. Pretend that the onion is the trunk of a tree; count the rings to find out the age of the tree. (4-6)

- 8. If you happen upon a tree trimming service trimming a tree, ask for a small cross section of a small limb. Use a magnifying glass to make discoveries about the make-up of the wood. Look for the age rings. During the X-mas season, tree sales stands often trim the ends of the trees trunks, ask for some samples. (2-6)

- 9. Find plants whose **stems** we eat. (Celery, asparagus, broccoli) (4-6)

- 10. Examine the **veins** on a leaf. Explain that veins are like hollow straws. Veins carry food, minerals, and water to and from the leaf. (4-6)

- 11. Make a **leaf place mat**. Place several leaves on paper and laminate. Make a second set of leaf cards to match. Learn the names of the trees your yard. During dinner play matching and identifying games. (2-6)

- 12. Use a magnifying glass to examine the leaf's surface. The surface is covered with small microscopic holes, called **stomata**. To show what the surface of the leaf might look like, cut a large green leaf from green construction paper. Use a straight pin to punch tiny holes all over the leaf, to represent the **stomata**. (4-6)

- 13. Examine the backs and the fronts of two different species of leaves. How are they alike? Different? Try using several different kinds of leaves. Trace your finger along the veins of the leaf. (2-6)

- 14. Are all of the leaves on a tree the same size? Guess? Randomly collect 10 leaves from the same tree and compare them. (3-6)

- 15. **Extract chlorophyll** from a leaf by crushing fresh green leaves and soaking them over night in nail polish remover (containing acetone). The chlorophyll will turn the liquid green. Help your child to make the **green color connection**. (4-6)

- 16. Discuss how trees and flowers in parks and woods get water? Discuss how trees must depend upon rain. ($2\frac{1}{2}$-6)

- 17. Plant seeds and keeps track of how many days it takes before the sprouts appear above the soil level. ($3\frac{1}{2}$-6)

- 18. Plant several seeds in soil. Push some seeds deeper into the soil. Will it take longer for the deeper seeds to emerge? Keep checking. (2-6)

- 19. Plant fast growing seeds such as radish and grass seeds. (2-6)

- 20. **Experiment**: Do seeds need to be in soil to grow? Sprout grass seeds on a sponge. Dampen a sponge; place in a bowl with a small amount of water. Sprinkle the sponge with grass seeds, keep the sponge damp. ($3\frac{1}{2}$-6)

21. Soak large dried lima beans overnight. Let your child carefully remove the thin outer skin (**seed coat**), to reveal the stored food. Gently open the two sides of the seed to reveal the small **embryo** safely tucked inside. The embryo is very small

and easily over looked. Examine with a magnifying glass. (2-6) Compare lima bean seeds to other seeds. (3½-6)

To observe the embryo, keep a few seeds covered with a damp paper towel. The embryo will begin to sprout.

- 22. **Experiment**: To observe how roots are formed from seeds, dampen several pieces of paper towel and place in a clear glass jar. Gently push lima bean seeds between the paper towel and the side of the jar. Make sure that the beans are clearly visible. Place the jar in a sunny spot. Check its progress. **The paper towel must be kept moist.** In a few days the beans should sprout. Describe what you see. These seeds may be planted in soil. (2-6)

- 23. Discuss how seeds travel from one place to the next. Take a walk in the woods or the park to find an odd tree within the midst of many of the same species. How do you think the seed landed here and grew into a tree? (3-6)

- 24. Take a walk through the neighborhood, woods, or park to look for seedlings. Seedlings are small plants that have begun to grow. Look near trees. Many seeds sprout and grow near the mother tree. (3-6)

- 25. Look around the floor of a moist wooded area to find moss growing. (2-6)

- 26. Plant a **sunflower**. These large flowers produce hundreds of seeds that can be easily observed as the seeds are developing on the face of the flower. (2-6)

- 27. Find a dandelion whose flower head has turned a puffy white. Examine the seeds; blow the soft white head of seeds. (2-6)

- 28. Collect seeds from different trees to examine and compare. (3-6)

- 29. Cantaloupes, honeydews, and watermelons have many seeds. Cut open a cantaloupe and a watermelon to compare the arrangement the seeds. Count the watermelon seeds.(2-6)

- 30. Use acorns as counters for number problems. There are many different species of acorns, see how many you can find? (2-6)

- 31. Some seeds develop in a pod. Discover some of these pods-Example-soybeans, okra, peas, green beans, snow peas, and Italian flat beans. Buy a sample of each to examine. Many super markets sell processed soybeans in the pods that are ready to eat. They are fun to eat and make an excellent snack. (2-6)

- 32. Take a walk in a field or woods to find **seeds pods**. Open the pods to examine the seeds. (2-6)

- 33. In the fall, take a walk to collect seeds from dead flowers. Save the seeds for planting next year. (2-6)

- 34. Sort and graph a package of mix nuts, by kinds, size, and color. (2-6)

- 35. Discover some plants whose **seeds** we eat. (Corn, beans, peas, nuts, sunflower) (2-6)

- 36. Examine cones from several different species of confers (evergreen) trees. (2-6)

- 37. If you happen upon a tree trimming service trimming a tree, ask for a small cross section of a limb. Use a magnifying glass to make discoveries about the make-up of the wood. Look for the age rings. During the X-mas season, tree sales stands often trim off the ends of tree trunks, ask for some samples. (2-6)
- 38. Which plant's **leaves** do we eat? (Cabbage, spinach, greens, herbs, lettuce) (2-6)

- 39. Trace leaves on cardboard and cut out. Use the cut-outs as stencils for fun projects. Learn the names (3-6)

- 40. In the spring, take a walk in the woods. Revisit the same woods in the summer and again in the fall and the winter, note changes. (2-6)

- 41. During the fall season, go for a ride in the mountains to see the changing colors of fall leaves. (2-6)

- 42. Walk in a pile of dried fall leaves, describe the sounds-crunchy, crackle, dry etc. (2-6)

- 43. In the fall of the year, find a deciduous tree whose leaves are beginning to change colors (loosing the green chlorophyll coloring). Try to find leaves that are in different stages of change. Examine the leaves. Note how some leaves have less green chlorophyll color. Put the leaves in sequential order-First place one with all of its chlorophyll and sequence the leaves down to one that has it bright fall color. (4-6)

- 44. Deprive a plant of water for a few days, it will begin to **wilt**. Basil plants are excellent for this project. Examine the leaves. Water the plant; keep checking to note how it recovers. (3-6)

- 45. Plant 2 pots of seeds. Feed one plant commercial plant food. Observe to see what happens. (2-6)

- 46. In the spring, keep a watchful eye to see when the trees and flowers in your yard start to bloom. Use a magnifying glass to examine the blossoms. Check the buds before, during, and after the blossoms begin to open. (2-6)

- 47. Collect the flowers of different trees. Discover how they are a like and how they are different. (2-6)

- 48. Examine a tulip or a lily to feel the **pollen** located on the **anther**. Rub the pollen on to your finger and examine with a magnifying glass. (2-6)

- 49. **Experiment**-To demonstrate how **sap** and water travel through the stems and veins of a leaf, place a fresh white carnation in water mixed with red food coloring. Leave over night. The red coloring will be sucked-up the stem and color the white carnation a pale red. Explain to your child that the stems of plants and trunks of trees have systems much like tiny straws that suck water and minerals from the ground. Give your child a glass of red juice and a straw. Let him/her pretend to be the stem of a flower sucking up water from the ground. (4-6)

- 50. Trees produce **sap** that is sweet and wet. The sweet sap from the sugar maple trees is collected, cooked, and appears on our store shelves in the form of **maple syrup**. Read books about how maple syrup is made. Sample pure maple syrup. Explain that he/she is eating food that the maple tree made for its own use. ($3\frac{1}{2}$-6)

- 51. Plant a **vine** plant, such as tomatoes, cantaloupes, pumpkins, cucumbers, beans, watermelons, or squash etc. (3-6) Go on a vine hunt.

- 52. Visit an aquatic garden or a pond. Look for plants that are growing in and around the water. "What did you discover?" (2-6)

- 53. Set-up a fish aquarium with live plants. Discover how plants supply oxygen to the water (2-6)

- 54. Some plants are poisonous.

Identify poison ivy; its 3-leaf cluster easily identifies this plant. (4-6)
Discover other poisonous plants.

- 55. Set-up a **terrarium**. (2-6)

- 56. Adapt a tree in your yard or near your home. Identify the tree. Feel the bark; collect its leaves and seeds. Visit the tree often to compare any changes, even in the winter. (2-6)

- 57. Compare the flowers of different kinds of trees. (2-6)

- 58. Discover plants whose flowers we eat. (Broccoli, cauliflower) (3-6)

- 59. **Touch** and **smell** a tree. Each tree gives off a different smell. A tree's smell can discourage animals from attacking. It is also the tree's smell that attracts insects to help in pollinating the tree's flowers. (2-6)

- 60. Compare the shapes of **conifer** evergreen trees to **broadleaf** trees. (4-6)

- 61. Brainstorm things that are made from trees. Find things around the house that are made from wood. (3-6)

- 55. After a walk in the woods, examine your clothing to see if any seeds have attached themselves to your clothing. These seeds are called **hitchhikers**? (3-6) Examine to see how the hitchhikers are attached to your clothing.

- 62. To observe the roots of a plant. Replant a potted plant. Use a magnifying glass to get a closer look. Find the **main root** and **small hair roots**. (3-6)

- 63. Take a walk in the woods or a housing construction sight to find a tree that has fallen. Examine the tree's exposed soft roots. Compare the soft roots to the hard bark covering the roots that can usually be seen appearing above ground. (3-6)

- 64. Trees have powerful roots. Take a walk to find places where the tree's roots have caused the heavy concrete sidewalk to be raised. (3-6)

- 65. Examine plants whose roots we eat. (potatoes, beets, carrots, radishes, parsnips, turnips) (2-6)

- 66. **Root a tree clipping**. Put a small clipping from a limb of a tree in a clear glass filled with water. Set the glass where it will get lots of sun. Keep a watchful eye to see what happens? Change the water often. Roots should begin to appear. (3-6)

- 67. **Sprout a sweet potato.** In a clear glass container, place the fat rounded end of a sweet potato. Stick toothpicks in the sides of the potato to prevent it from sinking. Place the potato into the container, letting the toothpicks support the weight of the potato. Add enough water to cover $\frac{1}{2}$ of the potato. Place the potato in sunny spot. Change the water often. In a few weeks the potato will begin to grow roots and leaves. (2-6)

- 68. Use a magnifying glass to examine **mulch**. Discover how mulch is made and how it is used. (3-6)

- 69. Cut a tulip bulb in half to observe the inside. Plant a **bulb plant**. (3-6)

- 70. Identify your state tree and flower. (4-6)

- 71. Identify your country's tree and flower. (4-6)

- 72. Discover miniature **bonsai trees**. (2-6)

- 73. Read about giant **redwood** trees. (3-6)

- 74. Look in the refrigerator, there is bound to be some green fuzzy **mold** growing on something. Use a magnifying glass to examine it. (3-6)

- 75. **Experiment**: To grow mold, moisten a piece of bread and seal it in a plastic bag. Place the bag on top of the microwave oven. In a few days, the heat from the microwave oven should produce mold. (3-6)

- 76. Compare the cores of an apple and a pear. (2-6)

- 77. **Cacti plants** are able to survive in extreme heat and with very little water. Plant a small cacti garden. (2-6)

- 78. In Arizona, the **Saguaro cacti** can grow to a height of 50 feet. Measure 50 feet. (4-6)

- 79. Many parts of the country have experienced **drought** and **flood** conditions. Find out what happens to trees and plants. (4-6)
- 80. Relax on the grass and sketch a tree. (4-6)

- 81. Plant herbs and use for cooking. (2-6)

- 82. Take a walk in the park or woods to look for **mushrooms**. Caution your child not to touch them. Many mushrooms are poisonous, even to the touch. (2-6) Note where it is growing.

- 83. Visit the store to collect a variety of mushrooms. Compare the mushrooms. (2-6)

- 84. After a hard frost, examine plants. (3-6)

- 85. Feel dew on plants. (2-6)

- 86. **Bamboo** is one of the fastest growing plants. It can grow over 15 inches a day. If you are lucky enough to live near a bamboo plant, get out and measure it. Mark one limb with a permanent marker. Check the next day. Bamboo is the giant panda bear's main food source. Learn more about these animals. (2-6)

- 87. Many large animals are **herbivores**. Learn about some of these animals. (**Elephants**, **hippos**, **giraffes**, and **gorillas** etc.) (3-6)

- 88. Discover some of the many animals that are **herbivores** (plant eaters). (3-6)

- 89. Read stories about animals that live in trees-Example-owls, squirrels, birds, monkeys, insects, and bats. (2-6)

- 90. Read stories about plants and seeds, "**Henny Penny, Jack and the Beanstalk**, and "**The Little Red Hen**." (2-6)

- 91. Check for signs that an animal has or is living in the tree? (2-6) Visit a pond or stream to find algae growing on rocks. Examine the leaves of a tree or a plant to find signs that an animal has been feeding. (2-6)

Water- Liquids (2-6)
Parent Background-
- The properties of water and other liquids can offer many fun science and math experiences.
- Liquids are apart of your child's daily experiences.
- There are always opportunities for incorporating rich descriptive liquid related vocabulary.
- The boundaries of a liquid must be contained.
- A liquid takes the shape of the container it is in.
- Liquids differ in their compositions. We will deal mainly with water.
Water
- All living things must have water.

Mary Taylor Overton

- The earth's natural resources exist in a closed ecosystem. Its water is constantly being recycled and used over and over again.
- Earth's plants and animals have been using the same water for millions of years.
- Water cannot be produced, children must be taught to look upon it as such.
- **It is a liquid.**
- It is **wet**.
- You can **pour** it.
- Takes the shape of the container
- It will **freeze** at 32 degrees F. and 0 Celsius.
- Will melt when temperature is above 32 F. or 0 Celsius and return to a liquid state
- Water will boil at 212 degree F.
- It **evaporates** as a **gas**.
- It **splashes**.
- It can cause things to **float**.
- Objects large and small will **sink** in it.
- It can be helpful as well as destructive.

Objectives-
1. To identify some of the properties of liquids
2. To identify water as a liquid
3. To learn related vocabulary
4. To observe water in everyday routines

Suggested Water Activities (2-6)
Children of all ages will have fun as they make discoveries about the properties of liquids-water. To accommodate all ages and levels most of the following activities are **open-ended**. Be creative!

Liquids-
Water is a **liquid**. It is wet and it takes the shape of the container into which it is in. Always keep talking and pointing out the properties of liquids. (2-6)
- "The milk that you are drinking is a **liquid**."
- "Before you can eat your canned fruit, I will **pour** off the **liquid**."
- "You must be careful, your soup is a **liquid** and it will **spill**."
- "The snow is **melting**; it will turn into a **liquid**?"

- 1. Discover things around the house that are **liquid**.
  (Caution-This is an excellent time to discuss the dangers of touching things that maybe hazardous.)

- 2. In the bathtub, provide different size and shape containers. Your child will make discoveries about **liquids** and **capacity**, as he/she **pours** water from one container into the next.

Water level-

Water level is the highest point of a liquid in a container.

- 1. **Experiment**-Just before running bath water let your child decide on the **water level**. "How high should we make the **water level**?" Mark the spot with a piece of colored tape. "You must keep a watchful eye, so that we can turn off the water when it reaches this **water level**."

- 2. Observe the water as the **water level rises**.

- 3. Use a clear measuring cup or plastic container to measure liquids. Example- For milk for cereal, "Pour enough milk to reach this level." For younger children ignore the numbers.

Displace-

When an object is placed into a liquid, its weight will cause the liquid to move (displace). Before your child gets into the tub, mark the **water level** with a piece of colored tape. When the child sits down, the weight of his/her body **displaces** the water. "When you sit down in the water, what will happen?" "Let's see how much the water will be **displaced**?" Mark the level. Use the vocabulary. "Stand up to see what happens?" (4-6)

- 1. **Experiment-1** If there are 2 children to take a bath, see what happens when the second child sits down. Mark with a different color tape. Have both children stand-up. What happened? (4-6)

- 2. **Experiment-2** For further exploration of **displacement**, fill a plastic container with water. To explore displacement give your child objects (heavy and light), to put into the water. Try using such things as a brick, cans of food of various sizes, anything with weight. (4-6)

- 3. **Experiment 3** Fill a glass almost to the top with water. See how much water is displaced when a cube of ice is added. Continue to add cubes to see what happens. Now take out the ice cubes. What happened? (4-6)

Sink or Float-

The properties of **sink** or **float** is based on the amount of water that is displaced (moved) by an object. An object will float if it displaces less water than its weight, and sink if it displaces more water than its weight. The density of an object also comes into play. A bar of soap or a large ship will not sink, because they are not completely dense (solid). Although they both appear to be solid, they have air pockets, making each less dense and weighing less than the water they are displacing. The same ship will sink when its air pockets fill with water and making the ship dense (solider-having no air pockets). A small penny is very dense and will quickly sink to the bottom of the container. The shape of an object will also determine whether it will sink or float. A wide, heavy object will float, because its weight is spread over a large area.

- 1. **Experiment 1**. While in the tub, discover the properties of **sink** or **float**. Provide a variety of tub toys. "Look that duck is **floating** on the water." (2-6)

- 2.**Experiment 2**. To demonstrate the meaning of **sink** and **float**, provide a large selection of objects to explore the properties of sink or float. Experiment with plastic, metal, wooden, and paper objects. Let your child select objects that will sink or float. Test them. (2-6)

- 3. **Experiment 3**. If an object is able to float try pushing down and holding the object under the water. If you take your hand away, what do you think will happen? What happened? (2-6)

- 4. Learn about large ships. If possible visit a location where large ships can be seen.

Drip and Drops- (2-6)
Allow the water from the shower, tub or sink to slowly drip. "Look, the water is **dripping**, let's clap and count the drips."
- 1. Mix water with food coloring. Use an eyedropper to **drop** water slowly on to paper towels or coffee filters. Try using several different colors. When the water hits the paper, it will spread out in different directions.

Spray-(2-6)
In the shower **sprays** water, use the word **spray** in your description of how the water is flowing.
- 1. **Experiment 1**. Most kitchen sinks have a sprayer for rinsing dishes. Let your child use the sprayer to rinse the dishes. (2-6)
- 2. **Experiment 2**. Fill an empty **spray** bottle with water and have fun spraying things, outside of course. (2-6)

Sprinkle-
- 1. Use your fingers to **sprinkle** water on your face.

- 2. Water flowers with a **sprinkling** can.

- 3. On a hot day, have fun running through the **water sprinkler.**

Water moves in a variety of ways, it **flows**, **runs**, **trickles**, or **ripples.**
The water **flows** down the **drain**, shower, wall, sink, or toilet.
- 1. "Count to see how long it takes the water to **flow** out of the tub?"

- 2. "Can you brush your teeth and get into your pajamas before the water **flows** from the tub?"

- 3. Rivers and streams flow. Visit a local stream to see the direction the water is **flowing.** (3-6)
- 4. Look at a map of the U. S. to find the Mississippi River, discover the number of rivers that flow into the mighty Mississippi River. (3-6)

Condensation-

Water is held in the air in the form of **water vapor.** Condensation occurs when warmer or cooler air comes into contact with a surface that is hotter or colder than the surrounding air. This causes the air on the surface of the object to change a liquid, **condensation.** This is because the cool air on the surface cannot hold the same amount of moisture as the surrounding warmer air. (4-6)

Condensation Experiments

- 1. **Experiment 1**-Put a tray of ice into a small metal pot. As the ice cools the pot, the warmer air directly touching the pot releases the water that it was holding. This results in water condensing on the outside of the pot. (3-6)

- 2.**Experiment 2**- Blow your breath on a mirror. What happened? Feel the mirror, what do you feel? See? (3-6)

- 3. **Experiment 3**-Observer the state of the bathroom. Check the mirror. Next turn on the shower. Make sure that the water is very hot and close the door. What do you think is going to happen? **Condensation** should occur. This should leave a **fog** on the mirror and other fixtures in the room. Return in a few minutes to see what has happened. (3-6)

- 4. **Experiment 4**- Fill a glass with ice and place it on a napkin. Later, check the napkin and the glass to see what is happening. (3-6)

Evaporation-

Water is found in three forms, **liquid**, **frozen** (ice), and a **gas** (steam). When water changes into a **gas** it is called **water vapor**. Water vapors then evaporate into the air. Water vapor in the air cannot be seen and young children have difficulty understanding that water can exist in the form of a **gas**. This is such an abstract concept and is better left alone, for the time being. When you mention gas to children, they immediately think of gasoline, which exists, in liquid form. Simply say, "The water **evaporated**-disappeared into the air." " Water is always in the air," and leave it at that. (3½-4)

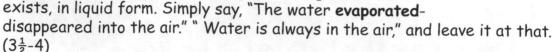

Water in the air is called humidity and is measured by the percentage of water being held in the air. Cold air holds less water than warm air.

- 1. **Experiment 1**-After a shower, when everything in the bathroom is wet; note places where water can be found. "What's going to happen to the water left on the shower tiles?" Come back later to find out what happened? "Where did the water go?" "What do you think happened?" "It **evaporated** into the air." (3-6)

- 2. **Experiment 2**- Spray a small amount of water on a mirror. Use a piece of paper to fan the water. Continue fanning until all the water has evaporated. (3-6)

- 3. **Experiment 3**- Wet 2 wash cloths with water. Put 1 inside of a plastic bag and hang-up the other. What do you think will happen? What did happen? What did you discover? (3-6)

- 4. **Experiment 4**- Put a small amount of water in 2 small pots. Cover one and leave the other uncovered. Bring both pots to a boil, until all of the water in the uncovered pot has **evaporated**. As the water is boiling note the steam that is escaping. What did you discover? Compare the amounts of water in each pot. (3-6)

- 5. **Experiment 5**- Boil water in a pot. This time hold the pot's top slightly above the steam. Catch some of the evaporating water. "What did you discover?" (3-6)

- 6. **Experiment 6**- To keep track of the rate of evaporation, put water into a clear plastic wide-mouth jar. Mark the water level. Check and mark again in 4 days and again mark the water level. Keep checking and marking until the water has completely evaporated. For a controlled jar use another jar just like the first. Add an equal amount of water as the first jar, and cover with the top. As this experiment evolves compare the 2 jars. (3-6)

- 7. **Experiment 7**- Wash a pair of socks. Tightly wring the water out of 1, and hang-up both socks to dry. See how long it takes for the water to evaporate from each sock. (3-6)

Rain- (2-6)

Water on earth is connected to the **earth's water cycle**. Rainwater vapors **evaporate** into the air in the form of a gas. Eventually the water is recycled and falls back to the ground as **rain, snow, sleet, frost, dew,** or **hail**. Children enjoy learning how the rain that is falling on them today is the same water that fell on the dinosaurs millions of years ago. Explain how water is **recycled** over and over.

- 1.**Observation**- After a rain, keep an eye on water puddles. Watch to see what happens to the puddles?

Absorb-(2-6)

Provide a sponge or paper towel. "Use this paper towel to clean up the spill. It will **absorb** all of the milk."

- 1. **Experiment 1**-Water a plant and watch the soil **absorb** the water.

- 2. **Experiment 2**- Use a dry sponge to **absorb** water in a small dish of water. What do you think will happen? What happened? This time use a sponge that has been soaked in water to clean up a spill. "What do you think will happen this time?

- 3. **Experiment 3**- To see which will or will not absorb liquids, try materials like-paper, various cloths, or foil. (3-6)

- 4. **Experiment 4**- Make French toast or bread pudding. Both require that the bread first be soaked in an egg mixture to absorb the liquid.

- 5. **Experiment 5**- Will sand absorb liquid? Put a small amount of sand into a container. Pour a small amount of water at a time over the sand, to see how much water will be absorbed.

- 6. **Experiment 6**-Add a small amount of milk to cereal, wait to see what happens. Did the cereal **absorb** the liquid?
- 7. **Experiment 7**- Cook rice for dinner. Observe how the water is absorbed

by the rice. Weigh the rice before and after cooking.

- 8. **Experiment 8**- Place raisins or prunes in water. What happened? Try using any dehydrated fruits, vegetable, or meat. (3-6)

- 9. **Experiment 9**-Place a small piece of cut fruit in a dish. Leave it uncovered, allowing it to dry-out completely-dehydrate-(this should take a few days). Check the fruit each day, to note its progress. After it has dehydrated completely, place it in water to see what happens. (3-6)

Freeze-
Water freezes when it is exposed to a **temperature** of 32 degrees F. or 0 centigrade or below.

- 1. Locate the 32 and 0 marks on a **thermometer**. (3-6)

- 2. Visit a pond or a puddle that has frozen. (2-6)

- 3. Plan a family outing to a local ice skating rink. (2-6)

- 4. Use a globe to locate the North and South Poles. Read books about these frozen lands. Identify **glaciers**, **icebergs**, and **ice caps**. (2-6) Look at a video about the poles.

- 5. **Experiment 1**-To show how icebergs float, add a tray of ice cubes or a large chunk of ice to bath the water. Also provide small toy boats. Have fun keeping the boats from hitting an iceberg. Read about the Titanic. (4-6)

- 6. **Experiment 2**- When the outside temperature is 32 degrees F. or below; put a small dish of water outside to freeze. Check each hour to see how it is progressing, note changes. (4-6)

- 7. **Experiment 3**- Freeze water in a large shallow pan. Have fun sliding objects made of different materials across the ice. (**Slick**, **slippery**, **smooth**, **glide**.) (2-6)

- 8. **Experiment 4**- Pour water into different shaped containers and place in the freezer. Make **frozen** treats in ice cube trays or paper cups. (2-6)

Melt-Defrost –
- 1. Take something from the freezer to **defrost**. To catch the excess water, place it on a plate. Keep checking its progress. (2-6)

- 2. **Experiment 1**-Freeze a bowl of water. Let it melt and **refreeze**. Feel

and describe the water at each of these stages. Think of words to describe how it looks and feels. (2-6)

Other Activities

- 1. Learn about weather that involves water-**Snow, floods, hurricanes, drought, hail, sleet, rain, frost, dew, thunderstorms, monsoons**

- 2. Drink using a straw. (2-6)

- 3. Visit a pond to see the plants that are growing in water and near the water. (2-6)

- 4. Visit a large waterway near your home. Use a map to locate it. (2-6)

- 5. Use various substances to see how well each dissolves in a liquid. Try dissolving sugar in both cold and warm tea to see what happens. (4-6)

- 6. Make lemonade; observe how the sugar **dissolves**. (2-6)

- 7. Combine water and oil together, shake to mix. What happened when you stopped shaking? Try making salad dressing. (2-6)

- 8. Discover how dams and underwater tunnels are constructed. (4-6) If there is tunnel or a dam nearby, visit it. Use a map to find tunnels that have been built under water and through mountains.

- 9. Use the globe to discover how much water covers the earth. Almost three fourths of the world is covered with water. (3-6)

- 10. **Water pollution** is a problem facing most waterways. Learn more about this problem. (4-6)

- 11. Discover the **earth's water cycle**. (5-6)

- 12. Find things that are **waterproof**. (3-6)

- 13. Discover animals that live in water. (2-6)

- 14. After a heavy rain find a sewer drain to see what happens to the excess water. Discover which rivers your sewer empties into. (3-6)

- 15. After a snowfall, measure 1 cup of snow. Mark the level and let the snow melt. Compare the amount of water left. (3-6)

- 17. Collect a cup of white show. Predicts what color the snow will be when it melts? (3-6)

- 18. Purchase a small rain gauge, they may be found in any dollar store. Use it to measure the amount of rain that falls. (5-6) Keep a record to compare measurements.

- 19. Make Jello. (2-6)

- 20. Make ice cream; observe how a **liquid** changes into a **solid**. (2-6)

- 21. Learn about sports that involve water. View videos of water sports- swimming, scuba diving, surfing, deep sea diving, skiing, fishing, and water skiing. (3-6)

- 22. Visit an ocean beach. Discuss how people and some animals cannot drink salt water. (4-6)

- 23. Make snow cones. (2-6)

Air-Wind (2-6)
Parent Background-
Oxygen and Wind

- The blanket of air (gasses) that surrounds the earth is called the earth's **atmosphere.**
- The atmosphere consists of several layers of air. Each layer has a different composition.
- The layer in which we live is called the **troposphere** and contains the air that we breathe.
- The earth's weather occurs in the troposphere layer.
- **Air-oxygen** is a gas that all-living things, both animals and plants must have. Without air-oxygen they would die.
- Contrary to belief, the air that we breathe is not made up of pure **oxygen.**
- More than $\frac{3}{4}$th of our oxygen is composed of the gas **nitrogen.**
- There are also a few other trace gasses in oxygen.
- The air that surrounds the earth is the same air that has surrounded it for millions of years.
- The earth is a **closed ecosystem.** All of our natural resources must be **recycled** to be used over and over again by plants and animals living on

earth.
- The green plants and animals living on earth have a partnership with the sun. All three are constantly recycling the earth's air supply.
- We cannot **see, smell**, or **taste air** yet it is all around us.
- It takes up **space**. This can be observed when we blowup a balloon.
- Air has **weight** and can be destructive. It can also be helpful as it gently moves and cools us on a hot day or helps to **pollinate** plants.
- It can generate electricity.
- The air that encases the earth is constantly moving. Because of the spherical shape of the earth, the **equator** receives more direct sun's heat than any other place on earth. Hot air from the equator spreads out across the globe. This hot air comes into contact with cold air spreading from the cold North and South Poles. The two conflicting air masses meet and compete for space, causing **wind** to form.
- Moving air is called **wind**.
- Wind patterns impact the weather all over the earth.
- Without oxygen, fire will not burn.

Objectives-
1. To understand some of the properties of air- wind
2. To understand that wind is moving air
3. To learn related vocabulary
4. To understand that all living things need air
5. To understand that air is a precious commodity that must be conserved

Vocabulary
- -Air
- -Deflate
- -Inflate
- -Windy
- -Strong
- -Turbulent
- -Gust
- -Breeze
- -Twirl
- -Pollution
- -Cycle

Suggested Air Activities-(2-6)
All of the following activities are appropriate for ages 2-6.
Always use Air-Wind Vocabulary
- 1. "The wind is **gusting**, look how hard it is **blowing**." "The wind is blowing a **gentle breeze**."

- 2. Instead of saying, "Blow up the balloon," say, " **inflate**." "Show me how you can inflate the balloon."

- 3. "Take a big breath by **inhaling** the air or oxygen." "Now **exhale**."

- 4. Talk about things you like to do on windy days.

- 5. Make windy day pictures. (3-6)

- 6. On a windy day, describe what is happening to the trees, flowers, animals, and people. How does it feel? Sound? (2-6)

- 7. On a windy day observe what happens to leaves on a tree. Leaves have the ability to hold to the branches, even in strong wind. Take a closer look to see how leaves are connected to the branches. (2-6)

- 8. Make paper airplanes. Try launching them from various heights. What happened? Change the angles of the wings to get more airlift. Get a book that shows how to construct more complicated paper planes. (2-6)

- 9. Use a blow dryer to dry your hair or anything that is wet. (2-6)

- 10. Wet your hands. How can you dry them without using a paper towel? Blow on your hands and wave them. (2-6)

- 11. In the restrooms of many stores, there are electric hand dryers. Use it to dry your hands. (2-6)

- 12. How do clothes dry in the clothes dryer? Allow your child to help load the dryer. Discover the location of the fan. (3-6)

- 13. Locate the fan in the dishwasher. (3-6)

- 14. After a rain, observe what happens to puddles when the wind blows. What do you see when the wind blows hard? What happens when it blows softly? (ripple) Come back later to check the puddle. Where did the water go? (2-6)

- 15. Buy an inexpensive **pinwheel**. Run fast and slow to make it turn. To observe the pinwheel as it turns in the wind put the stick the in the ground. (2-6)

- 16. Show your child how to cool hot soup or cocoa, by blowing it. (2-6)

- 17. Collect **dandelions**; **blow** off their white seed tops. Observe what happens. Follow a seed to see how far it goes. What do you think will happen when it lands? Plant a few of the seeds in a cup to see what happens? (2-6)

- 18. Collect maple tree seeds. These seeds have wings that are easily carried by the wind. Drop the seeds from high places to observe how they move. (2-6)

- 19. Discover that fire must have oxygen to burn. Extinguish a candle by placing a large glass over the candle. See how long it takes for the candle to go completely out. Discuss why the fire ceased. (4-6).

- 20. Run against the wind. How does it feel? Stand still to feel the wind as it blows against your face. Describe how it feels. (2-6)

- 21. Run with and against the wind. Is it harder or easier to run against or with the wind? (2-6)

- 22. Make a **windsock**. Attach 3 feet strips of ribbon to a piece of cardboard. Make a picture on both sides of the cardboard. Attach a string on the opposite end. Hang the windsock so that it can catch the wind. (2-6)

- 23. Hang a wind chime where it can catch a **breeze**. (2-6)

- 24. On a windy day observe how the wind blows things around. "The wind is **gusting** or **blowing** hard." "What happened when the wind stops?" (2-6)

- 25. When the wind is blowing, place a piece of paper on the ground. Try catching the paper. What happened? (2-6)

- 26. Keep an eye out for **weather vanes**. What happened when the wind blows? (3-6) Buy a weather vane.

- 27. Catch wind in a plastic bag. Use an empty plastic grocery bag, the kind with handles for holding. Hold one end as you run. Try using small and large bags. What happened? How can we trap the wind inside of the bag? (2-6)

- 28. Inflate a small plastic or paper bag. Pop the bag. (3-6)

- 29. On a cold day, blow your breath into the cold air. What do you see? Go inside and breathe. What happened this time? (2-6)

- 30. On a windy day, wet two pieces of cloth. Hang one outside in the wind and the other inside. What do you think will happen? (2-6)

- 31. Drop small drops of water on the table. Use straws to blow drops across the table. (2-6)

- 32.   Open a soda can and listen to the sound of the air escaping. What is making that sound? (2-6)

- 33. Roll-up a piece of paper into a horn. Try making different **sounds**. (2-6)

- 34. Buy a small plastic horn and have fun blowing it. (2-6)

- 35. Purchase paper party horn blowers the kind that rolls and unrolls. What makes it unroll and return? (2-6)

- 36. Learn about **wind instruments**. (4-6)

- 37. **Make a kite** to fly. Purchase a kite kit. (2-6)

- 38. Pretend to be a kite flying in a **strong wind**, a **gentle breeze** or a **strong gust** etc. (2-6)

- 39. **Make a sailboat**. Stick a straw through a small Styrofoam meat tray. Attach a paper sail to the straw. Use the sailboat in the bathtub. How can we make the boat move? Blow on the sail to move the boat. (2-6)

- 40. **Howl** like the wind. (3-6)

- 41. Drop pebbles into the puddle to see the water ripple. Experiment with large and small pebbles.(2-6)

- 42. Pretend to be a large bird-**soar, glide**- in the wind (2-6)

- 43. Do you think that birds can fly on windy days? Go outside on a windy day to look for birds. What did you find out? (4-6)

- 44. Blow **bubbles**. Fill a bowl with water and add dish detergent. Use straws to blow bubbles. Observe what happens as you blow air into the water? (2-6)

- 45. Purchase commercial bubbles. (2-6)

- 46. Make a cloth flag and hang it outside in the wind. (2-6)

- 47. Slowly **inflate** a balloon. Squeeze the balloon to feel the air pressure pushing against the sides. (3-6)

- 48. Inflate a balloon and let it go. What do you think will happen? What made it fly around the room? (3-6)

- 49. **Inflate** a beach ball or any plastic inflatable toy. (3-6)

- 50. Inflate a balloon, hold it under water and let the air escape. What happened? (4-6)

- 51. Blow small pieces of down feathers. Down feather can be found in old down pillows or coats. (2-6)

- 52. Secure a long piece of crape paper ribbon to a pencil. Hold the pencil and run. Make a second. Try making large and small circles. (3-6)

- 53. Use a straw to pick up small pieces of tissue paper. Suck hard enough to pick up the paper. (3-6)

- 54. Make a **parachute**. Cut a piece of cloth 12x12 inches and attach a string to each of the 4 corners. Tie the other end of the strings to a small toy play figure. Drop the parachute from high places. View a video of parachute jumping. (3-6)

- 55. Decorate a paper plate and use it as a **flying saucer**. Try different sized plates. (3-6)
- 56. Learn about **windmills**. (4-6)

- 57. Fish are able to breathe underwater by means of gills. Learn how fish breathe. (4-6)

- 58. Read books about **air pollution**. (4-6)

- 59. Read books about hot air balloons. If possible go to a hot air balloon show. (2-6) Look at videos of hot air balloons.

- 60. Fold a piece of paper into a fan. Let your child make a picture on the paper and then fold. Use the fan to move air. What do you feel? Why? What happens when you fan slowly? Fast? (2-6)

Mary Taylor Overton

Vocabulary Building

Creating Life-Long Math and Science Learners helps the child to develop a strong **vocabulary foundation**. The following are some suggested vocabulary that parents can incorporate into their child's daily activities. It is not necessary for your child learn every word.

- Introduce words within the context of related experiences; this gives your child a reason for learning and using the word.
- Do not drill a list of isolated words.
- Children must hear and use a word over and over again, before it becomes a part of their working vocabulary.
- **If you use the words, so will your child.**
- Most of the words can be applied to more situations, an example of this is the word **waterproof**.
- **Waterproof-**
- "Use the umbrella it is **waterproof**."
- "The bark on a tree is **waterproof**."
- "I can swim with my watch, because it is **waterproof**."
- "This plastic bag will keep your sandwich from getting wet, it is **waterproof**."
- "Your mittens are **waterproof**."
- "Our tent is **waterproof**."
- "A turtle's shell is **waterproof**."
- "The shower curtain is **waterproof**."
- "The plastic tablecloth is **waterproof**."
- "The bird's feathers are **waterproof**."

Activities-
- Find things around the house that are **waterproof**.
- Have fun in the bathtub exploring various items to determine if they are **waterproof**.

*Words inside ( ) are the opposite meanings

A
- **Above**-(Below)
- **Absorption**-To become apart of by soaking it up
- **Abstract**-A concept that can not be seen
- **Aestivation**-To hibernate during the day to escape the sun's heat
- **Add**-To join together
- **After**-Behind (before)
- **Air**-The blanket of atmosphere

226

surrounding the earth

- **Air pollution**-Foreign trash that has been released into the air by man
- **All**-Total
- **Almost**-Nearly
- **Amount**-The total
- **Amphibians**-Frogs, salamanders, toad, newts- cold blooded; begin life on earth in water; vertebrates; hatch from eggs
- **Angle**-The point at which 2 lines intersect; meet
- **Annuals**-Plants that grow once and die
- **Answer-To respond**
- **Antarctic-The South Pole (North Pole)**
- **Antennae**-Feelers; located usually on the head
- **Apart-Separate (together)**
- **Arachnid**-spiders, ticks
- **Arctic-North Pole (South Pole)**
- **Area-Any space with boundaries**
- **Are there more? - (Are there less?)**
- **Arrange-To put things in order**
- **Assist-To help**
- **Asymmetrical- An object that has unequal sides (symmetrical)**
- **Atmosphere-The blanket of air that covers the Earth**
- **Attract**-A force that brings objects together (Repel)
- **Autumn**-Fall, a season

B

- **Backbone**-Vertebrae- (invertebrates)
- **Bacteria**-Small particles found on everything-causes decay and disease
- **Balance**-To be equal
- **Bare trees**-Trees without leaves

Words inside ( ) are the opposite meanings

- **Bark**-The hard outer protective cover on trees
- **Bask**-To rest in the sun to warm
- **Beak**-The mouth of a bird
- **Beat**-To hit
- **Behavior**-Action
- **Behind**-(Front)
- **Belong**-Go-together (does not belong)
- **Below**-Under (above)
- **Between**-(Middle)
- **Big**-Bigger, biggest (little)

- **Birds**–Animals with feathers; hatch from eggs; warm blooded; vertebrates
- **Birth**–Emerge into the world
- **Bitter**-A taste identified by the sensors on the tongue
- **Blank**-Empty
- **Blizzard**-A heavy snow fall
- **Blood**-A body fluid that carries oxygen, nutrients and waste throughout the body
- **Blossom**-Buds that open
- **Body**-A structure that holds the part of animals and plants
- **Boil**-Water boils at 212 degrees F.
- **Bones**-Skeleton
- **Born**-To enter the world
- **Bottom**-(Top)
- **Brain**-The body's computer
- **Breakdown**-To separate into smallest units
- **Breakfast**-The first meal of the day
- **Breathe**-To exhale and inhale air
- **Branches**-Part of a tree
- **Breakdown**-To divide into smaller pieces
- **Breeze**-A soft calm wind
- **Bright**-Brighter- brightest
- **Buds**-The small protective cover of a leaf or flower before it blossoms
- **Build**-To construct

*Words inside ( ) are the opposite meanings
- **Burrow**-A hole dug by an animal in the ground **Butterfly**-Invertebrates; insects

C
- **Calendar**-An instrument used to show and measure time- months, days, year
- **Camouflage**- To hide in the surroundings
- **Capacity**-The amount that a container will hold
- **Capture**-To grab
- **Carbon dioxide**-A substance given off by humans and animals as they breathe.
- **Carnivore**-Animals and plants that are meat eaters
- **Caterpillar**-The larva stage of a moth or butterfly
- **Center**-Middle
- **Change**-To alter
- **Chain reaction**-A series of events linked
- **Chip**-A small piece from a larger object
- **Chlorophyll**-A green substance found in leaves that absorbs the sun's light energy. Green plants use this energy power to make food.

- **Choose**-Select
- **Chrysalis**-A casing spun by a caterpillar, while inside it changes into an adult butterfly.
- **Circle**-A round continuous connecting line; edges equal from the center
- **Circular**-Having a circle shape
- **Circular motion**-To turn around and around
- **Clear**- Light is able to pass through(Transparent)
- **Climate**-Average weather conditions over a period of time
- **Clockwise**-To count 1-12/ Counter clockwise-(counterclockwise)
- **Close-(**Open)
- **Cocoon-**A casing spun by a caterpillar, inside it changes into adult moth
- **Column**-To put in a line or row

* Words inside ( ) are the opposite meanings
- **Coins**-A metal chip with an assigned value
- **Cold**-A state of being (hot)
- **Cold blooded**-Animals that cannot control their inner body temperature
- **Collect**-To gather
- **Colorless**-Not having color
- **Combine**-Mix
- **Compare**-To evaluate 2 things
- **Compass**-An instrument to show direction
- **Compound eyes**-An eye made- up of many lens; invertebrates have compound eyes
- **Conclusion**-To decide; end
- **Condensation**-Water droplets, caused by hot and cold air mixing
- **Cone**-A 3 D shape; a holder for seeds of conifer trees (evergreens)
- **Connect**-Attach
- **Coniferous**-Trees that grow seeds in cones
- **Construct**-To build
- **Consume**-To eat or use
- **Container**-An object that holds things
- **Copy**-Duplicate
- **Core**-The center, inside
- **Cost**-Amount to be paid
- **Count**-To match 1-to-1;
- **Count backwards**- 10,9,8,7,6,5,4,3,2,1,0
- **Counter clockwise**-To count backwards the opposite direction of the hands on a clock-(**clockwise**)

- **Cube**-3 D shape; a solid square
- **Cup**-8 ounces
- **Curve**-Bend
- **Cycle**-A reoccurring order
- **Cylinder**-A 3 D shape: Both top and bottom are circular: curved sides

D

- **Daily**-Occurs everyday
- **Damp**-Slightly moist

* Words inside ( ) are the opposite meanings

- **Day**-(Night)
- **Day light**-Light during the day (moonlight)
- **Dark**-Darker- darkest
- **Death**-The lack of life
- **Decay**-Decompose
- **Deciduous**-Broadleaf trees whose leaves turn colors and fall and are dropped
- **Decompose**-A process by which the tissues of animals and plants are broken down
- **Defrost**-to melt (freeze)
- **Dehydrate**-Remove by drying out the liquid contents
- **Demonstrate**-To show how
- **Desert**-A barren dry land
- **Describe**-To tell about
- **Dew**-Droplets of water found on cars or grass in the morning
- **Different**- Not alike (Same)
- **Direction**-Step-by-step how to
- **Discover**-To find
- **Dissolve**-To become apart of
- **Diurnal**-Active during the day (nocturnal)
- **Divide**-Separate (whole)
- **Dorsal fin**-The fin located on the backs of some fish
- **Dollar**-A unit of value in our money system equal to 100 pennies
- **Down**-A position (up)
- **Down feathers**-Soft feathers close to the skin
- **Drip**-A slow leak
- **Drought**-Lack of water (flood)
- **Droplets**-Small drops of a liquid
- **Dry**-To evaporate (wet)
- **Dry up**-To evaporate
- **Dull**-Not sharp; duller-dullest (shine)
- **Dust**-Small bits of dirt

* Words inside ( ) are the opposite meanings

E

- **Ears**-An organ used to gather sound waves (Sense of hearing)
- **Earth**-One of nine planets-Third planet from the sun
- **Echo**-A vibration of sound bouncing off of an object
- **Echolocation**-Sound waves that bounce off an object
- **Ecosystem**-Natural surrounds that animals and plants interact for survival of each
- **Edge**-Outline
- **Elliptical**-An oval shape
- **Embryo**-A small plant or animal developing; before it is born
- **Emerge**-To come out of
- **Enemy**-An opponent (friendly
- **Energy**-Power
- **Environment**-Surroundings
- **Equal**-Same
- **Equator**-An imaginary line that divides the Northern and Southern Hemispheres
- **Estimate**-Guess
- **Evaporate**-When a liquid turns into a gas
- **Even**-Equal
- **Evening**-Early night (approx. after 6 P. M., before 12 midnight)
- **Evergreen**-Plants whose leaves are green all year around
- **Exhale**-(Inhale) To breathe out
- **Expand**-Enlarge
- **Experiment**-To try
- **Explain**-To tell about the process
- **Explore**-Examine; to learn about
- **Eye**-The organ that gathers light images; Sense organ- seeing

F

- **Fall**-The season after summer; autumn
- **Fangs**-Hollow front teeth of a snake; dispenses poisons
- **Far**-Farther-farthest (Close)
- **Fat**-Oily
- **Feel**-Sense organ- touch

* Words inside ( ) are the opposite meanings

- **Female**-A girl; woman (Male)
- **Few**-(Many)
- **Fill**-A point of capacity
- **Find**-(Loose)
- **Fins**- Bony exterior parts of a fish's body, uses in movement

- **Firm**-(Loose)
- **First**-Front (last)
- **Follow directions**-To use instructions
- **Flat**-Not having height
- **Float**-(Sink)-An object that does not displace its weight in a liquid
- **Flood**-(Drought) Too much water to be absorbed or held
- **Flow**-A steady stream
- **Flow of energy**-Energy that begins with the sun's energy. The energy is used by green plants to make its food-animals eat the plants; the energy is transferred into their body-energy is transferred from prey to predator
- **Flower**-The part of the plant that produces seeds
- **Food chain**-Energy is past between the sun, plants and animals living together
- **Food web**-A interconnected sharing of available food sources by animals living in the same area
- **Forecast**-Predict
- **Forest**-An ecosystem
- **Forest floor**-The ground
- **Forward**-(Backward)
- **Frame of reference**-Knowledge about a certain topic
- **Fresh**-New
- **Freeze**-Water freezes at 32 degrees F.
- **Front**-(Back)
- **Frost**-A light freeze
- **Fruit**-Part of the plant containing seeds
- **Full**-(Empty)

Words inside ( ) are the opposite meanings

- **Fur**-A covering on some mammals

G

- **Gallon**-A liquid measurement; 4 quarts
- **Gather**-To put together
- **Geometry**-Study of space, lines and shapes
- **Germinate**-The point at which seeds begin to grow
- **Gills**-Fish have gills for breathing
- **Go**-(Stop)
- **Graph**-A compiled picture story of information
- **Group**-To put things together
- **Grow**-Enlarge; to multiply
- **Guess**-Estimate

H

- **Habitat**-Natural surroundings
- **Hair**-A body covering found on some mammals
- **Half**-Equally divide
- **Harden**-(Soften)
- **Healthy**-Not sick
- **Heat**-Warmth
- **Heavy**-Heavier, heaviest (light)
- **Height**-Measurement of how tall
- **Hemisphere**-The earth is divided into 2 halves-Northern and Southern; separated by the equator
- **Herbivore**-Animals that eat plants and vegetables
- **Hibernate**-To rest or sleep for the winter
- **Hide**-Cover
- **High**-(low)
- **Hold**-Support
- **Hollow**-Empty inside-Like a straw
- **Hoof**-A kind of a foot found on some mammals
- **Hot**- Something holding heat(Cold)
- **Hour hand**-A part of a clock used to show a unit of time
- **Hour**-A measurement of time, 60 minutes
- **Human**-People-mammals

Words inside ( ) are the opposite meanings

I

- **Ice**-Icy-frozen water
- **Icicles**-Long frozen ice drippings
- **Idea**-A thought
- **Identify**-To know; point out
- **Imagination**-To think about in ones mind; create
- **Inch**-Smallest unit of line measurement
- **Incline**-Slight straight rise in the surface
- **Inflate**-To expand (Deflate)
- **Inhale**-To breathe in (Exhale)
- **Insectivore**-Animals and plants that eats only insects
- **Insects**-Invertebrates- 3 body parts, 6 legs; compound eyes; cold-blooded; hatch from eggs
- **Inside**-(Outside)
- **Interesting**-Something you like to engage in
- **Invent**-To make something new
- **Invertebrates**-Animals without a backbone (vertebrates)
- **Invisible**-Cannot be seen (visible)

J

- **Join**-To put together
- **Joint**-Connects 2 bones to make bending and movement possible

K

- **Kingdom**- All living things are classified as members of the plant kingdom or animal kingdoms.

L

- **Landscape**-The surrounding land
- **Layer**-To place one top of the other
- **Large**-Larger-largest (Small)
- **Last**- (first)
- **Leap year**-Occurs every 4 years-one extra day-366 days
- **Leaves**-The part of the plant that makes food
- **Left**-(Right)
- **Length**-(Width)-Measurement to tell how long; distance
- **Less**-(More)

Words inside ( ) are the opposite meanings

- **Lift**-To pick up
- **Light**-Lighter-lightest (dark)
- **Lighting**-A fast strike
- **Limb**-Arms on animals; branches on trees
- **Liquid**- A state; takes the shape of the container; water is a liquid (Solid)
- **List**-To put together in an organized state

- **Live**-Having life (Dead) (nonliving)
- **Living Things**-Animals and plants-breathe, reproduce, eat, grow, and must have water
- **Log**-A part of a tree
- **Long**-Longer-longest (short)**Loud**-(Soft)
- **Low**-(High)
- **Lungs**-Two organ in the chest area that absorbs oxygen from the air, releasing carbon dioxide into the air

M

- **Magnify**-To enlarge (reduce)
- **Male**-Boys, men (female)
- **Main root**-The major root of a plant
- **Mammals**-Animals that have hair or fur, females produce milk,

backbones, breathe air, live births
- **Many**-More than 1
- **Map**-Directions
- **Markings**-Patterns on animals or plants
- **Match**-To group things that are alike
- **Material**-The physical make-up of an object
- **Measure**-To collect information about size
- **Meat**-The flesh of animals
- **Melt**-To be in a liquid state (Freeze)
- **Mercury**-A planet; a liquid in a thermometer
- **Middle**-Center
- **Migrate**-To move from one place to another
- **Minutes**-A measurement of time- clock time- 60 seconds equal 1 minute
- **Mix**-To combine together
- **Moisture**-Water

Words inside ( ) are the opposite meanings
- **Molt**-To shed ones skin
- **Months of the Year**-12 months- January-December)
- **Moon**-A natural satellite
- **Moonlight**-Light from the moon
- **More**-Uneven (Less)
- **Morning**-Time between daybreak and noon
- **Most**-More than $\frac{1}{2}$
- **Mother Nature**-A nickname to indicate the natural events of the world
- **Motion**-Movement
- **Mountain**-A tall land mass, created millions of year ago
- **Movement**-Motion

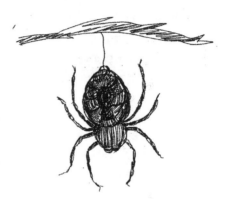

N
- **Name**-An identification label
- **Narrow**-(Wide)
- **Natural**-A state of being unchanged; pure (manmade)
- **Near**-Nearer- nearest (far)
- **Nectar**-A sweet wet food produced by flowers; (sap in trees)
- **Never**-(Always)
- **New**-(Old)
- **None**-Not any
- **Nocturnal**-animals that are active during the evening or night (diurnal)
- **Non-living**-Something not having life
- **Non-poisonous**-Not harmful (poisonous)

Mary Taylor Overton

- **Noon**-12 o'clock in the day; lunch time
- **Not as much**-(Less than)
- **Normal**-The agreed upon criteria
- **Numbers**- Symbols and the values assigned to each-

O

- **Objectives**- What the outcome should be
- **Ocean**-A large body of salt water
- **Ocean floor**-The bottom of the ocean
- **Off**-(On)
- **Old**-(new)
- **Omnivores**-Animals that eat meat and plants
- **Open**-(Close)

Words inside ( ) are the opposite meanings

- **Opposite**-Completely different
- **Orbit**-To encircle; go around
- **Order**-To sequence; to put into its place
- **Ordinal numbers**- 1st, 2nd
- **Over**-(Under)
- **Ounces**-Smallest unit of weight measurement; 16= to 1 pound
- **Out**-(In
- **Outer**-(Inner)
- **Oxygen**-A gas (Air)

P

- **Paper**-A process of making sheets from wood pulp
- **Parasites**-Animals or plants that get their nutrition by living off other plants or animals
- **Part**-(Whole)
- **Pattern**-To repeat; a design
- **Paw**-A type of foot of some mammals
- **Pebble**-A small rock
- **Pectoral fin**-Fins on the sides of fish; used for steering
- **Pentagon**-A 5 sided shape; 5 corners
- **Photosynthesis**-A process whereby plants use the sun's energy, chlorophyll, carbon dioxide, water and minerals to make food for the plant
- **Pit**-A hard protective container that encases a seed; a deep hole
- **Planet**-Mercury, Venus, Earth, Mars, Jupiter, Saturn, Uranus, Neptune, Pluto; circle the sun
- **Plant**-A living thing that is not animal
- **Pod**-A seed holder
- **Point**-A spot

- **Poisonous**-Dangerous and sometimes deadly (non-poisonous)
- **Pound**-A measurement of mass- 16 ounces
- **Problem**-A task to solve
- **Pulley**-A machine used to lift things; uses ropes and wheels
- **Pyramid**-A 3 D shape; base is a square or a rectangle; sides triangle; meet together at the top called a vertex
- **Predict**-A guess; based on ones knowledge
- **Predator**-The hunter; (prey)

Words inside ( ) are the opposite meanings

- **Preen**-Birds use their beaks to groom their feathers
- **Prepare**-To make ready
- **Pressure**-The resistance exerted on an object
- **Prey**-To be hunted for food
- **Primary Colors**-Red, blue, yellow colors, all other colors are made from these 3 colors.
- **Project**-A task
- **Pull**-(Push)
- **Pump**-To make liquid flow
- **Push**-(Pull)

Q

- **Quart**-A liquid unit of measurement- 32 ounces
- **Quarter**-A U.S. coin with a value of 25 pennies; $\frac{1}{4}$ of the whole
- **Question**-To ask; inquire
- **Quick**-Fast (slow)
- **Quill**-A part of a feather

R

- **Rain**-Water vapor that condenses and falls to the ground as droplets.
- **Rainbow**-Reflection of colors through rain vapor in the atmosphere
- **Rain gauge**-An instrument used to measure rain
- **Rainfall**-Rain that is falling
- **Rain water**-Water collected from falling rainfall
- **Receptors**-Sensitive nerve endings that send messages from the body's senses to the brain.
- **Rectangle**-A shape; 4 corners; 2 equal long sides; 2 equal sides
- **Rectangle Prism**-A 3 D rectangle shape
- **Recycle**-To use again
- **Reflection**-Light waves bouncing off of an object
- **Refreeze**-To turn a liquid back into a solid

- **Release**-To let go
- **Reptiles**-Snakes, turtles, lizards, crocodiles, alligators; cold-blooded; vertebrates; most hatch from eggs
- **Right**-A position (Left)
- **Ripe**-Fully grown; ready to be eaten
- **Rock**-A hard stony material
- **Roll**-To turn
- **Rot**-Decay
- **Rote counting**-numbers in sequence-1,2,3,4,5,6,7-
- **Rotate**-To turn
- **Rough**-(Smooth)

S

- **Salty**-Having salt taste
- **Salt water**- Water containing salt
- **Same**-(Different)
- **Salmonella**-A sometimes deadly bacteria found in eggs or raw poultry
- **Sap**-Sweet liquid produced by trees
- **Scales**-A measuring tool
- **Scatter**-To place randomly
- **School of fish**-Many small fish living together
- **Seasons**-Predictable weather patterns-**Summer, fall, winter, spring**
- **Seed**-A container holding an embryo and a food supply
- **Seedling**-(A young plant)
- **Seed coat**- The outer protective layer of a seed
- **Seed pod**-A casing that a seed develops
- **Semi-circles**-1/2 circle
- **Senses**-5 senses- hearing, seeing, tasting, smelling, feeling
- **Sequence**-Order of events
- **Shades of color**-The same color in varying degrees of lightness and darkness
- **Shadow**-An obstruction of light by an object, causing a darkness of the object to be cast
- **Shapes**-A space defined by lines
- **Shallow**-Not deep
- **Sharp**-(Dull)
- **Shell**-The outer layer

Words inside ( ) are the opposite meanings
- **Shinny**-Light bounces off; a bright light
- **Short**-Shorter, shortest/tall
- **Sink**-(Float)
- **Size**-Measurement
- **Skeleton**-A frame of bones
- **Skin**-The largest organ in the body; contains sensory receptors; sense

of feel
- **Skinny**-(Fat)
- **Slow**-Slower, slowest (Fast)
- **Small**-Smaller, smallest (large)
- **Smell**-Sense -smell
- **Smooth**-(Rough)
- **Snow**-Small flakes of frozen water
- **Soft**-(Hard)
- **Soil**-Dirt
- **Solar system**-The sun and the 9 planets of which the earth is one; everything in space
- **Solid**-Not hollow; no space in-between
- **Solve**-To find the answer
- **Sour**-A taste detected by taste buds in the mouth and on the tongue
- **Space**-Area
- **Species**-A sub-division within the plant and animal kingdoms
- **Spin**-To turn
- **Speed**-A measurement of how fast an object is traveling
- **Sphere**-See geometry section- A solid circle.
- **Split**-To separate
- **Spotted**-Having dots
- **Spring**-The season after winter
- **Square**-A shape 4 equal sides ;4 corners
- **Standard measuring tools**- Tools used to indicate an amount-rulers, calendars, clocks, money, thermometers
- **Steam**-Water in a state of gas
- **Sticky**-Gooey
- **Stream**-A small flowing body of water
- **Striped**-Lines
- **Stump**-A smaller portion
- **Subtract**-Take away from

Words inside ( ) are the opposite meanings
- **Sugary**-A sweet taste
- **Summer**-The season after spring
- **Sun**-A fireball of gas
- **Sunset**-Just before dark
- **Sunshine**-Light from the sun
- **Surface**-The top layer
- **Survive**-To last
- **Sweet**-(sour)

T

- **Take away**-To remove
- **Tall**-Taller, tallest (short)

- **Tally**-To add together; total
- **Tame**-(Wild)
- **Taste**-Sense organ; mouth and tongue
- **Temperature**-A measurement of hot or cold
- **Thermometer** -An instrument used to measure degrees; tells how hot or cold
- **Thick**-Thicker, thickest
- **Thin**-Thinner, thinnest
- **Thorax**-The middle section of an insect's body
- **Tight**-Tighter, tightest (loose)
- **Today, tomorrow, yesterday**
- **Tool**- An object that does work
- **Top**-(Bottom)
- **Touch**-Sense of feel (skin)
- **Tough**-(hard)
- **Translucent**-An object that very little light can pass through
- **Transparent**-Light is able to pass through
- **Triangle**-A shape; 3 corners; 3 sides; top meet at vertex
- **True**-(False)

U

- **Under**-(Above)
- **Underground**-Beneath the surface
- **Underneath**-Below
- **Underwater**-Below water

Words inside ( ) are the opposite meanings

- **Universe**-All things in space

V

- **Vegetables**-Plant material
- **Vertebra**-Small bones that make up a backbone
- **Vertebrates**-Animals that have backbones
- **Vibrate**-To move in a pattern
- **Vine**-A long climbing stem; flowers, fruits and vegetables sprout from one stem
- **Volcano**-A large bulging crack in the earth's crust that pours hot molten rocks
- **Volume**-The amount that a container will hold

W

- **Warm**-Warmer, warmest (cool)
- **Warm-blooded**-Animals that can keep their body temperature constant; Humans 98.6

- **Water**-A liquid; necessity for living things; living things-plants and animals-are composed of large amounts of water
- **Waterproof**-Protection against getting wet
- **Wavy**-not smooth,
- **Weather**-The climate
- **Weatherman**-Someone that studies and predicts weather
- **Web**-A structure spun by spiders
- **Weight**-The density of an object; the mass
- **Wide**-Wide, widest/narrow

Y

- **Yard**-A unit of measurement equal to 36 inches
- **Year**-A measure of time- 365 days; leap year 366 days every four years
- **Yesterday**-The day before today
- **Young**- (old)

Words inside ( ) are the opposite meanings

Fun with Action Words

Capitalize on young children's fascination with descriptive words, by using the following words to describe the actions of animals.

➤ Try using and dramatizing the following action words.
➤ Brainstorm other animals that engage in the same actions.

Read stories about animals that your child cannot identify.

- Fly-Like a bird
- Kick-Like a kangaroo
- Howl-Like a wolf
- Buck-Like a horse
- Hibernate-Like a groundhog
- Molt-Like a snake
- Scamper-Like a squirrel
- Hatch-Like a bird
- Gallop-Like a horse
- Walk –Like a spider
- Jump- Like a cricket
- Hop-Like a bunny
- Swoop-Down like an owl
- Slitter- Like a snake
- Slide-Like an alligator
- Creep-Like a cat
- Crawl-Like a bug
- Hang-Like a bat
- Swim-Like a fish
- Hunt-Like fox
- Snap- Like a crocodile
- Weave-Like a spider (spin)
- Strike- Like a snake
- Spit-Like a lama
- Waddle-Like a duck
- Work-Like bees
- Stalk-Like a tiger
- Run-Like a cheetah
- Sting-Like a wasp
- Bellow-Like an elephant
- Roar-Like a lion
- Sing-Like a bird
- Chirp-Like a cricket
- Hiss-Like a snake
- Migrate-Like a whale

- Leap –Like a frog
- Trot-Like a horse
- Climb-Like a monkey
- Bark-Like a dog
- Purr- Like a kitten
- Crow-Like a crow
- Soar-Like an eagle
- Pinch-Like a crab or lobster
- Dive-Like a porpoise
- Chatter-Like a monkey
- Peck-Like a chicken
- Burrow-Like a groundhog
- Butt-Like a goat
- Camouflage-Like an insect
- Dig-Like an earthworm
- Build- Like a beaver
- Swing-Like a monkey
- Growl-Like a dog
- Stand-Like a flamingo
- Scratch-Like an ape
- Squeeze-Like boa constrictor
- Eat-Like a pig
- Sleep-Like a kitten
- Yell-Like a monkey
- Slow-Like a snail
- Look-Like a hawk
- Hide-Like an animal
- Clap-Like a seal
- Waddle- Like a penguin
- Coo-Like a pigeon
- Breathe-Like a fish
- Stalk-Like a lion
- Roam-Like an animal
- Walk-Like an elephant
- Swell-Like a blow fish